D1153425

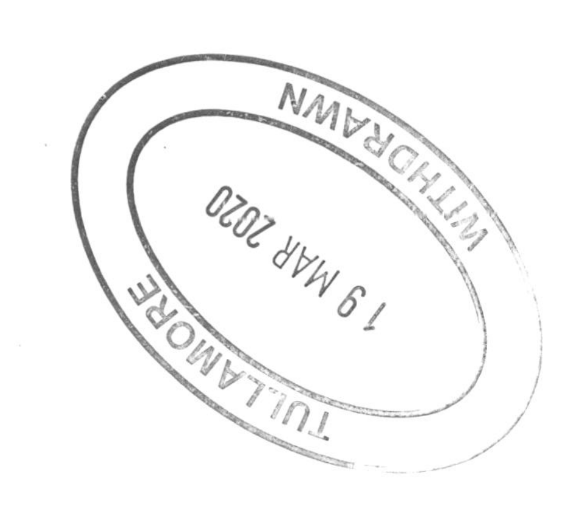

The Dawn of INDUSTRY

1750 TO 1810

Published by The Reader's Digest Association Limited
London • New York • Sydney • Montreal

Laudanum.--POISON

Contents

A bridge to the future

The elegant iron bridge over the River Severn in Shropshire was the first cast-ion bridge erected anywhere in the world. Construction began in 1777 and the bridge opened in 1780. Today it sits in a setting of rural serenity, but in the 18th century nearby Coalbrookdale – where the iron structure of the bridge was forged – blazed with coke-fuelled iron foundries.

Introduction

By the mid-1700s humanism and rationalism, the dominant ideologies of the preceding decades, had opened up the world to scientific investigation. The shift in focus from the sacred to the secular brought with it a gradual but radical transformation of society's values. In political and social terms, the defining moment of the period was the French Revolution. Shockwaves from the upheaval were felt across the whole of Europe. Before long, another revolution would begin in industry and fundamentally reshape the world.

Intellectual endeavour now seemed capable of rising to any challenge, as advances in theoretical science were converted almost instantly into practical applications. The disciplines of astronomy and optics began to probe the realms of the unimaginably large and infinitesimally small. Machines in the workplace brought technology directly into people's daily lives. Steam was the driving force behind the Industrial Revolution, but scientists were already probing the secrets of the power that would replace it – electricity.

In the turmoil that ushered in the 19th century, knowledge and power began to converge. It was in the name of extending Russia's influence and prowess that Catherine the Great promoted the arts and sciences, cultivating the leading thinkers of the age. Napoleon's Egyptian campaign deployed an army of intellectuals to survey Egypt past and present, with a clear nationalistic agenda. Henceforth, no colonial expedition was complete without its scientists. In turn, learning and research took on a moral dimension. There was universal assent in the early industrial age that the rise of science and technology would improve he lot of humanity, both materially and spiritually. This unquestioning faith in progress would prove long-lived.

The editors

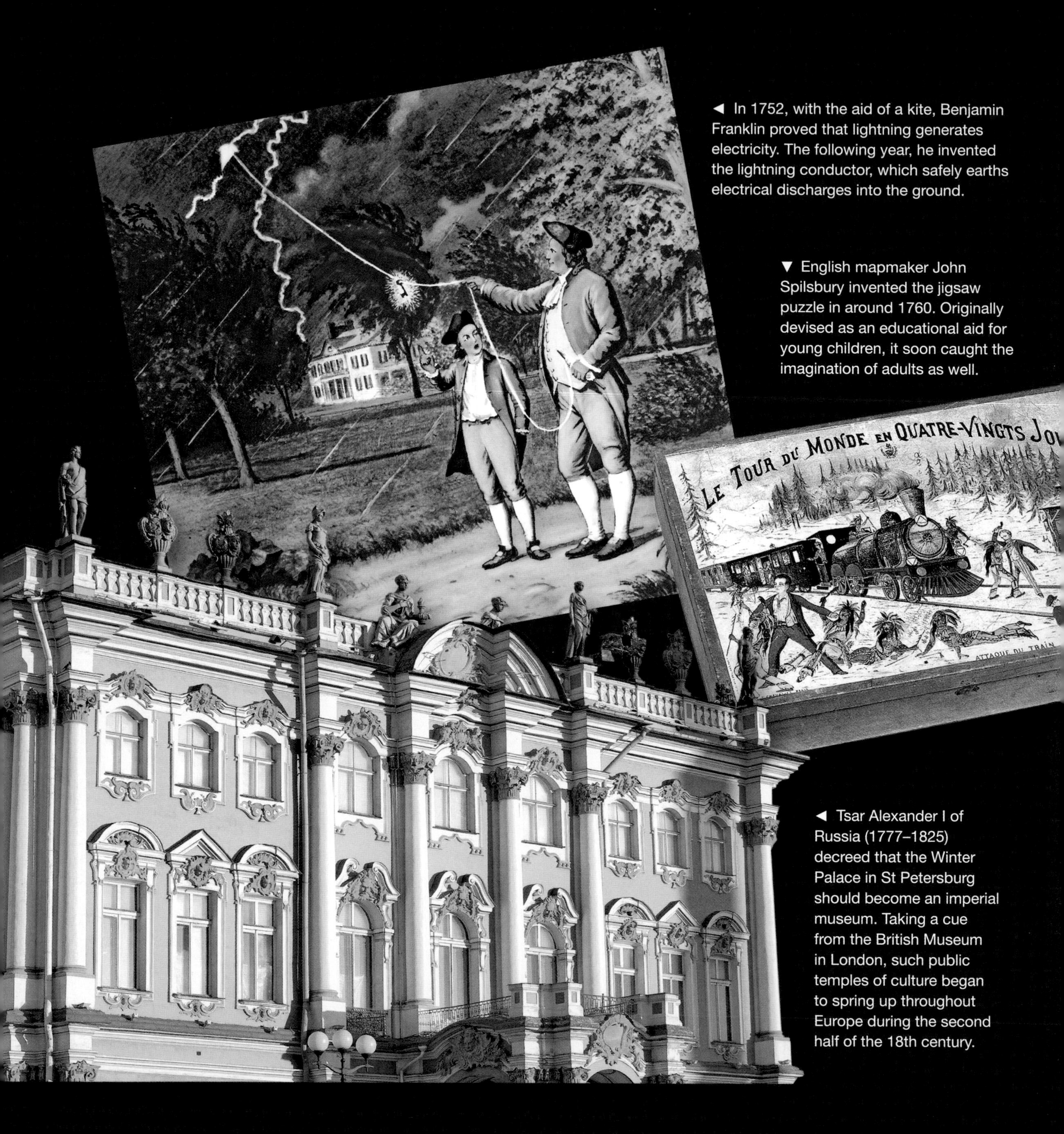

◄ In 1752, with the aid of a kite, Benjamin Franklin proved that lightning generates electricity. The following year, he invented the lightning conductor, which safely earths electrical discharges into the ground.

▼ English mapmaker John Spilsbury invented the jigsaw puzzle in around 1760. Originally devised as an educational aid for young children, it soon caught the imagination of adults as well.

◄ Tsar Alexander I of Russia (1777–1825) decreed that the Winter Palace in St Petersburg should become an imperial museum. Taking a cue from the British Museum in London, such public temples of culture began to spring up throughout Europe during the second half of the 18th century.

In 1750, the French philosopher Denis Diderot was appointed as general editor of the French *Encyclopédie*, and one of the key works of the Enlightenment got underway. Inspired by the 17th-century English scientist Francis Bacon, and working alongside the

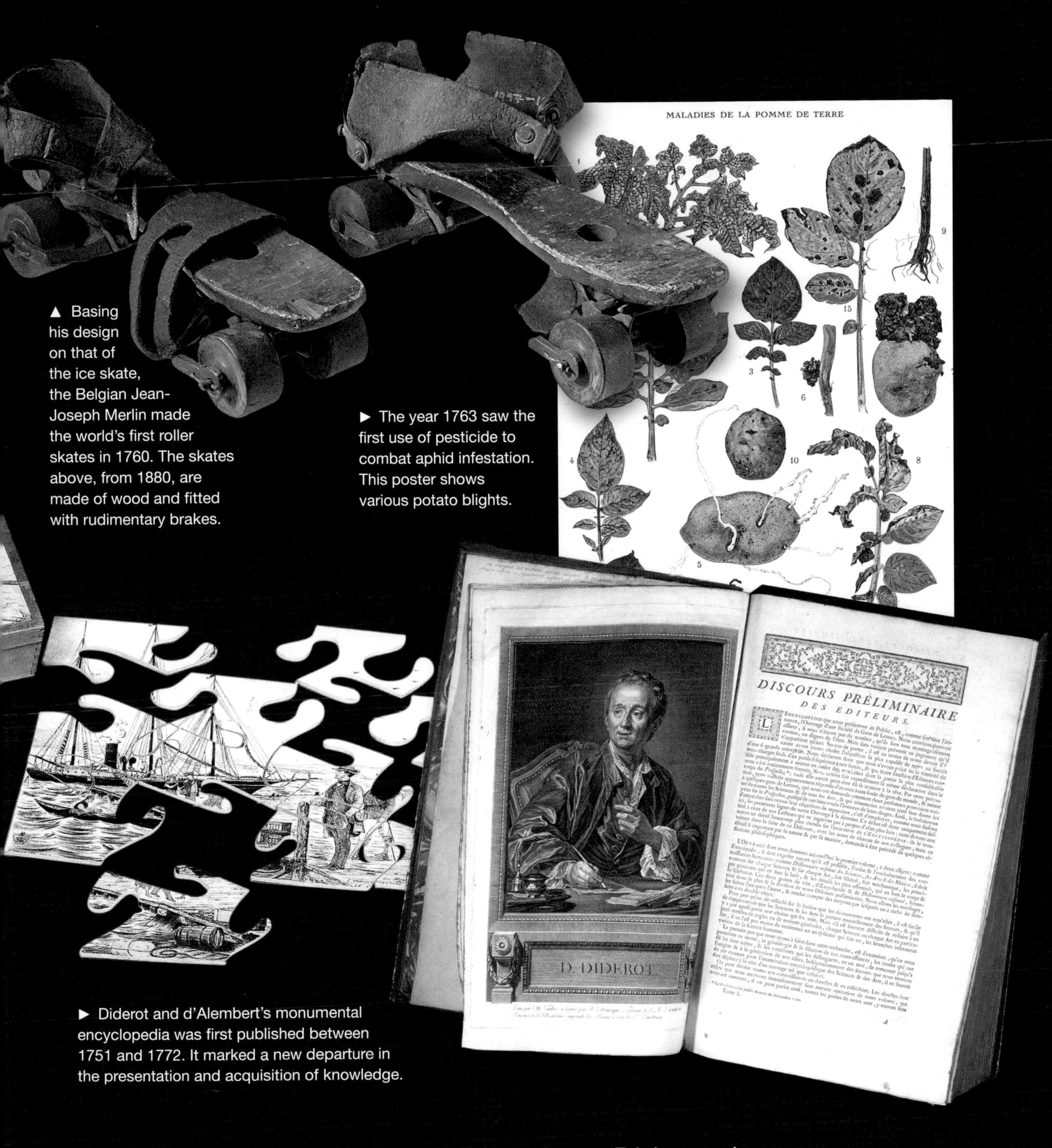

▲ Basing his design on that of the ice skate, the Belgian Jean-Joseph Merlin made the world's first roller skates in 1760. The skates above, from 1880, are made of wood and fitted with rudimentary brakes.

► The year 1763 saw the first use of pesticide to combat aphid infestation. This poster shows various potato blights.

► Diderot and d'Alembert's monumental encyclopedia was first published between 1751 and 1772. It marked a new departure in the presentation and acquisition of knowledge.

mathematician Jean le Rond d'Alembert, Diderot drew up a prospectus for a 'systematic dictionary of the sciences, arts and crafts'. The end result would be a monumental work, published in 35 volumes, celebrating the scientific, technological and artistic

► In 1774 Joseph Priestley, the father of chemistry, isolated oxygen but failed to recognise the importance of his discovery. He also did pioneering work on photosynthesis in plants.

▼ On 23 November, 1770, a 'steam carriage' built by Nicolas-Joseph Cugnot ran out of control and knocked down a wall at the Paris Arsenal. Brakes were being used on coal trucks by this time, but had not been fitted to Cugnot's monster vehicle.

▲ In 1768 the entrepreneur Philip Astley created the modern circus, when he began staging equestrian extravaganzas near Westminster Bridge in London. The shows were initially performed in the open air, but they soon moved into a lavish purpose-built arena.

achievements of the age. In gathering together this vast range of human learning, the editors were aiming for nothing less than to change 'the way people think'. It was no coincidence that public museums and scientific collections also came into being at this

► A calorimeter dating from 1785 that belonged to French chemist Antoine Lavoisier, who established the fundamental principles of modern chemistry.

▼ Erected in 1779, the iron bridge at Coalbrookdale, was the first metal bridge in the world. It was built by the ironmaster Abraham Darby III, and opened up an entirely new use for the cast iron from his foundries.

◄ Meteorology came into its own from the late 18th century as scientists began investigating atmospheric phenomena. The first weather station was set up in Germany in 1780. This anemometer for measuring wind speed (left) dates from 1875.

time. The period from the mid-18th to early 19th century was one of extraordinary intellectual dynamism, hallmarked by the rise of the new discipline of chemistry. Under the guidance of pioneers such as Joseph Priestley and Antoine Lavoisier, chemistry

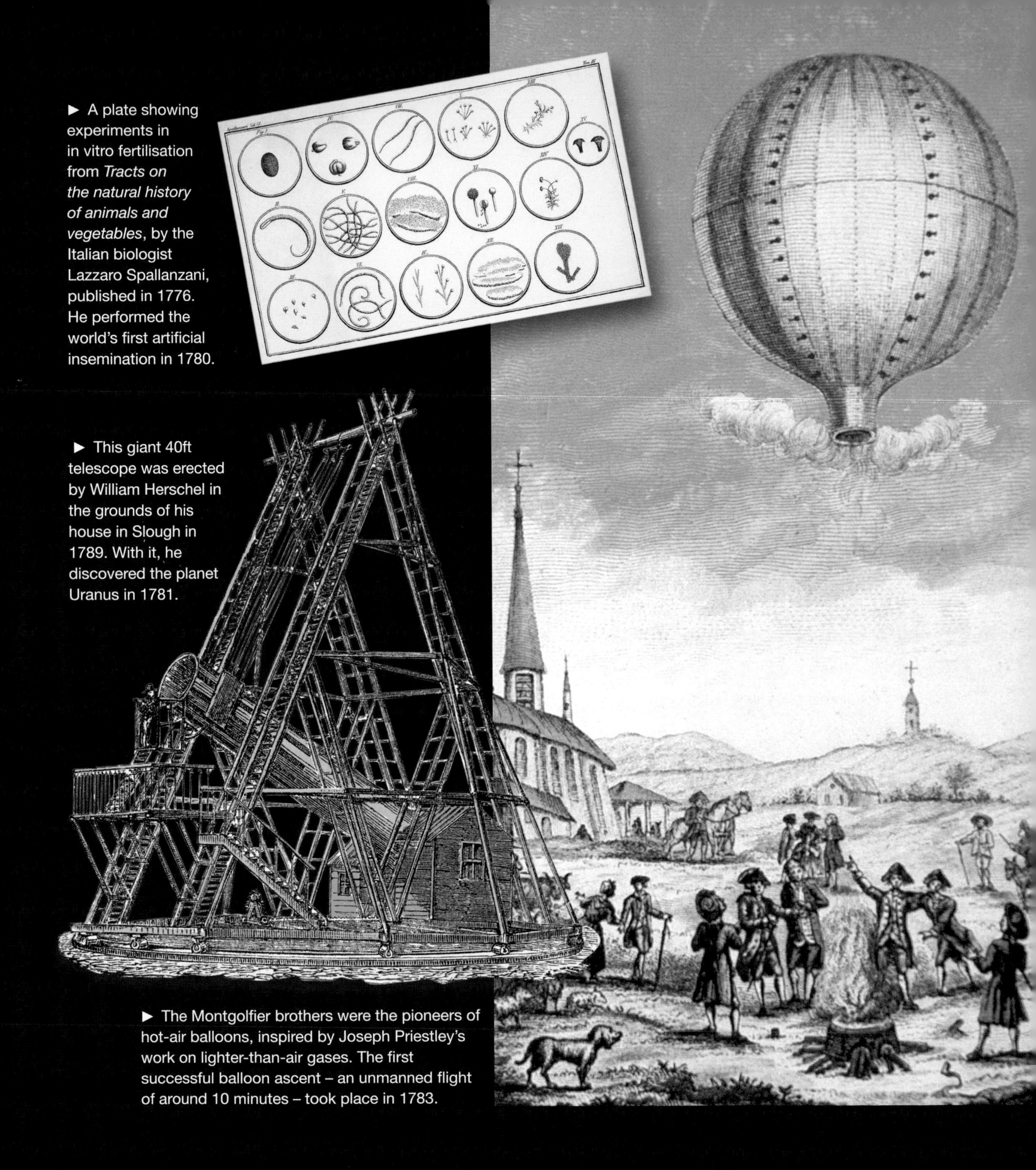

► A plate showing experiments in in vitro fertilisation from *Tracts on the natural history of animals and vegetables*, by the Italian biologist Lazzaro Spallanzani, published in 1776. He performed the world's first artificial insemination in 1780.

► This giant 40ft telescope was erected by William Herschel in the grounds of his house in Slough in 1789. With it, he discovered the planet Uranus in 1781.

► The Montgolfier brothers were the pioneers of hot-air balloons, inspired by Joseph Priestley's work on lighter-than-air gases. The first successful balloon ascent – an unmanned flight of around 10 minutes – took place in 1783.

shrugged off its mystical origins in alchemy and became an empirical science. Other areas of scientific enquiry also began to carve out their niches: in the life sciences zoology, botany, biology, physiology, pharmacology and medicine provided new insights into

◄ A model of the steam-and-sail-powered *Savannah*, which made the first steam-assisted crossing of the Atlantic in 1818. The first prototype steamship was successfully tested by French inventor Jouffroy d'Abbans in 1783.

▲ The guillotine was used for the first time in 1792. It became the quintessential symbol of the 'Reign of Terror' during the French Revolution.

▲ In 1792, two French astronomers began measuring the meridian arc from Dunkirk to Barcelona. Eventually, their work would yield the definitive unit of the metre as the basis for a new system of weights and measures.

realms as diverse as artificial insemination and the alleviation of pain. It was above all the practical application of theoretical knowledge that fired the public imagination. Innovations like pesticides increased crop yields; inoculation against smallpox

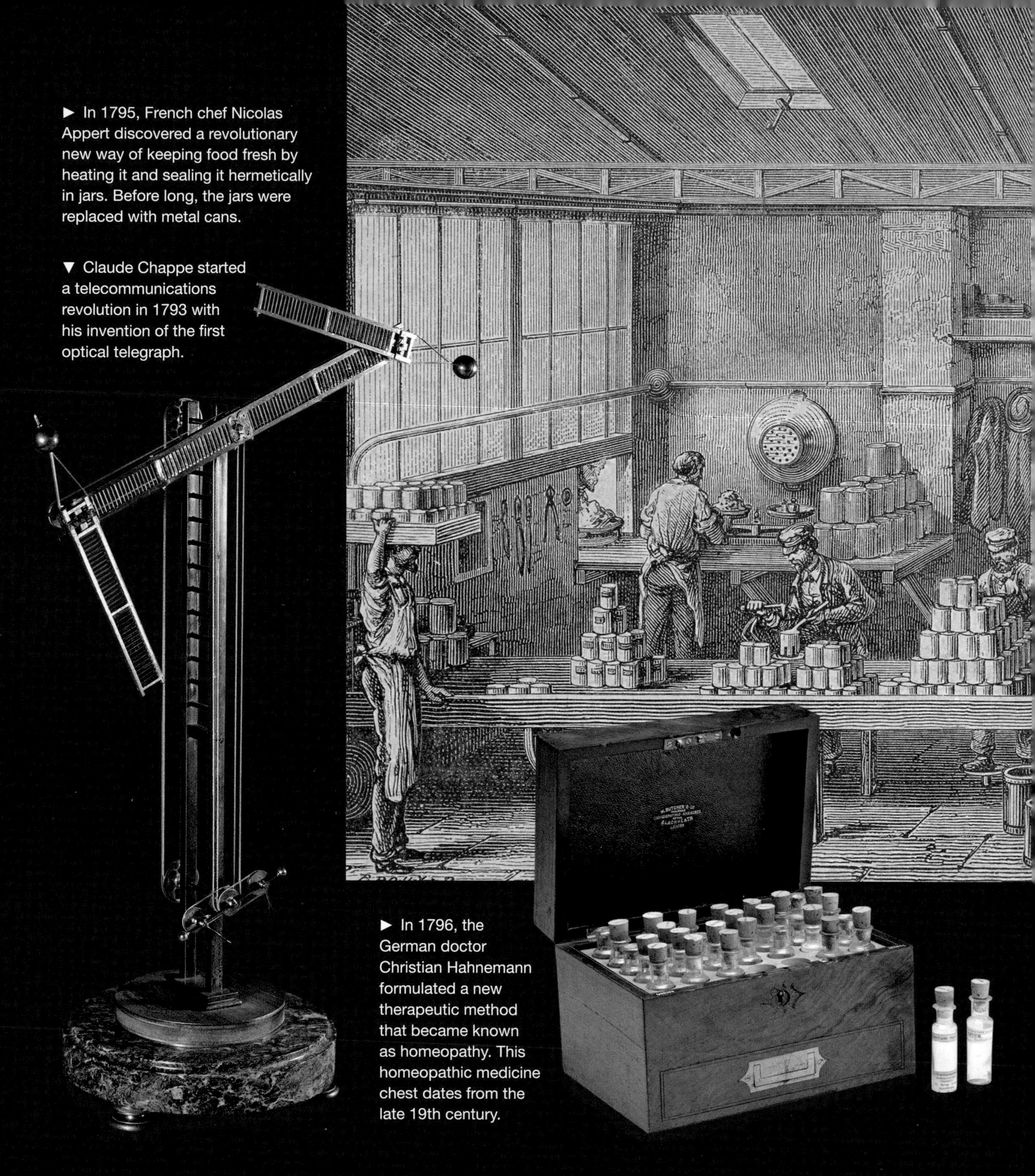

► In 1795, French chef Nicolas Appert discovered a revolutionary new way of keeping food fresh by heating it and sealing it hermetically in jars. Before long, the jars were replaced with metal cans.

▼ Claude Chappe started a telecommunications revolution in 1793 with his invention of the first optical telegraph.

► In 1796, the German doctor Christian Hahnemann formulated a new therapeutic method that became known as homeopathy. This homeopathic medicine chest dates from the late 19th century.

safeguarded against a killer disease; the installation of gas lighting in city streets began to disperse the urban gloom; the lightning rod, Benjamin Franklin's simple but effective device, gave peace of mind in violent electrical storms. The turn of the 19th century saw an

◄ A needle once used by Dr Edward Jenner to inoculate patients against smallpox in a process known as variolisation. In 1796, Jenner pioneered a more effective method: inoculation with cowpox, or vaccination.

▲ In the great Age of Exploration, European navigators Bougainville and Cook captained expeditions to the Pacific, discovering and charting unknown lands. Scholars and scientists went along to study and catalogue new fauna and flora.

◄ In 1796, Alois Senefelder invented lithography, a printing process that did not involve engraving. This poster by Fernand Louis Gottlob was produced by the lithographic process.

early hint of the coming age of long-distance communication in the optical telegraph. Prototype steamboats showed that it was possible to travel on water without the aid of sails or oars, although for the time being maritime exploration was still conducted under sail.

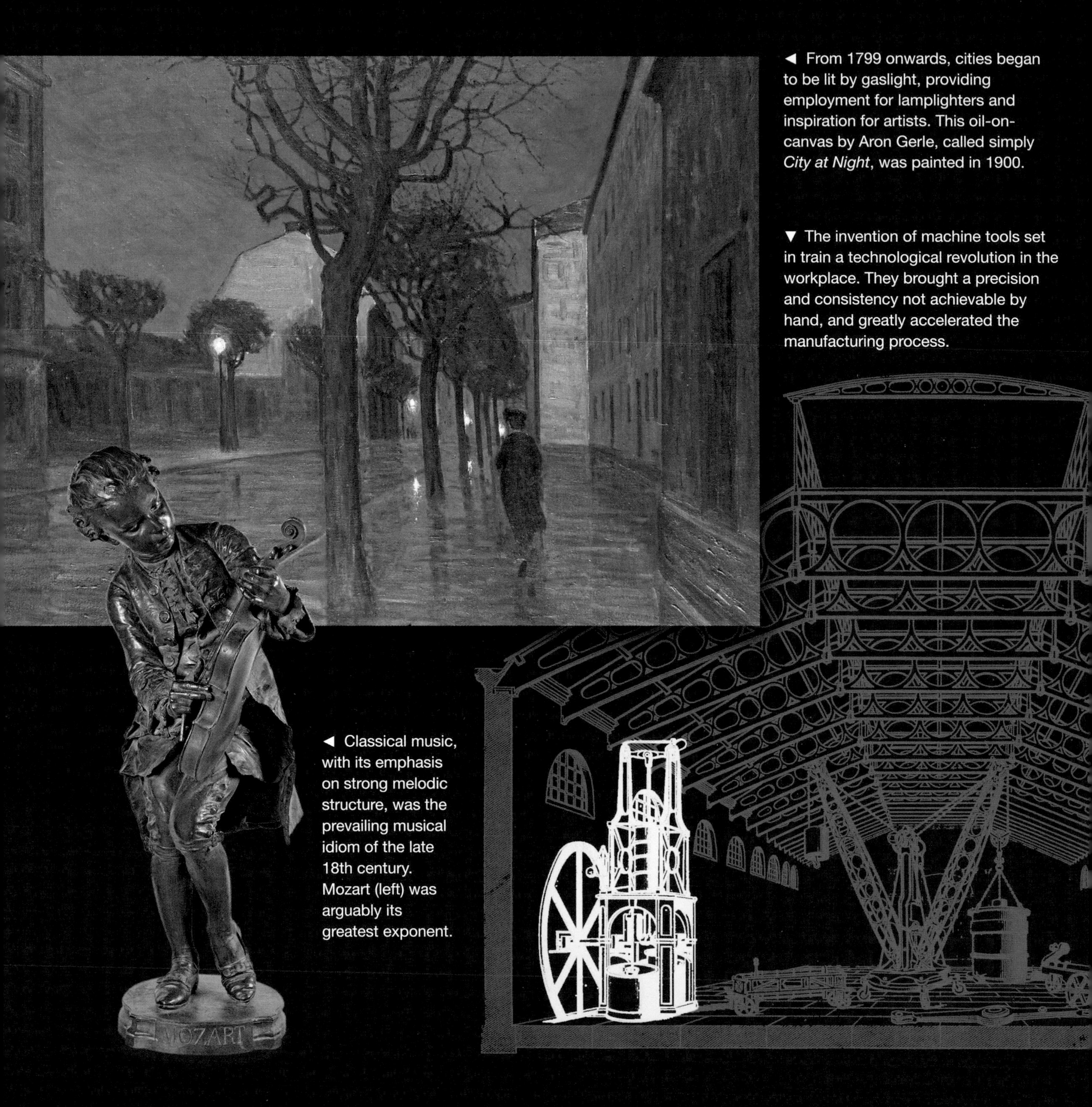

◄ From 1799 onwards, cities began to be lit by gaslight, providing employment for lamplighters and inspiration for artists. This oil-on-canvas by Aron Gerle, called simply *City at Night*, was painted in 1900.

▼ The invention of machine tools set in train a technological revolution in the workplace. They brought a precision and consistency not achievable by hand, and greatly accelerated the manufacturing process.

◄ Classical music, with its emphasis on strong melodic structure, was the prevailing musical idiom of the late 18th century. Mozart (left) was arguably its greatest exponent.

James Cook and Louis-Antoine de Bougainville charted the oceans, discovering unknown worlds, while Humboldt made inroads into the vast tropical forests of Latin America. But the most radical change of pace in this period was in working lives. Inventions like the

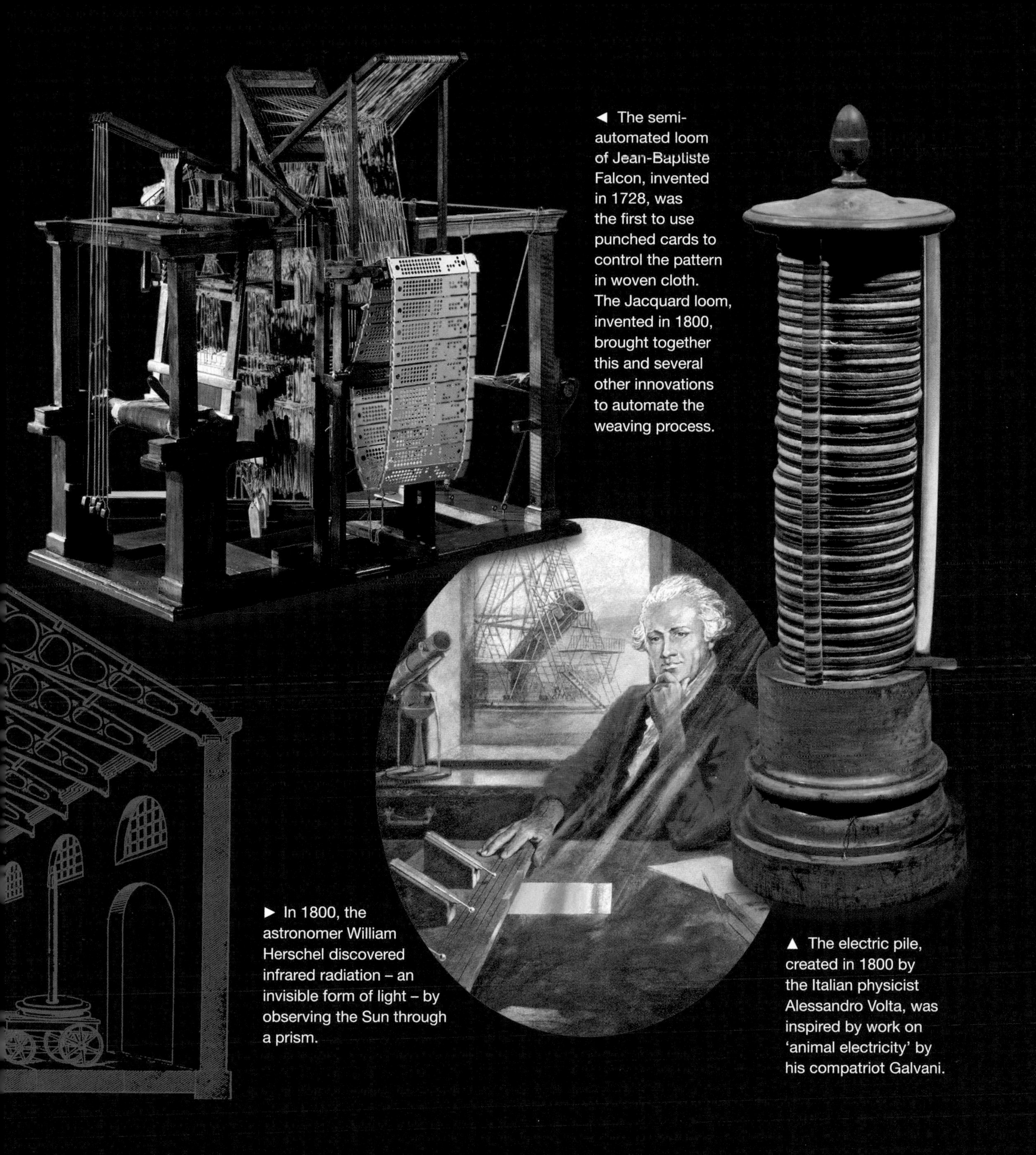

◄ The semi-automated loom of Jean-Baptiste Falcon, invented in 1728, was the first to use punched cards to control the pattern in woven cloth. The Jacquard loom, invented in 1800, brought together this and several other innovations to automate the weaving process.

► In 1800, the astronomer William Herschel discovered infrared radiation – an invisible form of light – by observing the Sun through a prism.

▲ The electric pile, created in 1800 by the Italian physicist Alessandro Volta, was inspired by work on 'animal electricity' by his compatriot Galvani.

Jacquard loom and Maudslay's lathe boosted output and streamlined the manufacturing process, making large sections of the workforce redundant, while forcing those who remained to become ever more productive. The economic system of industrial

▼ By around 1800, advances in chemistry had enabled scientists to isolate morphine, the active ingredient of opium, which had a powerful analgesic effect. Laudanum, a tincture of opium, was widely used in the 19th century.

◄ In around 1800, the invention of an effective chromatic trumpet finally enabled trumpet-players to produce the full range of notes.

▲ Napoleon's Egyptian campaign was a military disaster, but the scientific and archaeological team he took with him scored some notable successes. In particular it lead to the discovery of the Rosetta Stone, which enabled ancient Egyptian hieroglyphs to be deciphered.

capitalism was in the making, starting a process that still continues today: the depopulation of the countryside and growth of cities. By 1800, more and more people were being drawn to urban areas to find employment: the population of London, then the biggest

▲ From 1806, the introduction of the Congreve rocket gave warfare a new dimension in firepower. These early weapons are now regarded as the ancestors of modern space rockets.

► Electroplating was invented in 1805 by Luigi Brugnatelli. Largely ignored at the time, the process is now widely used in the manufacture of computer chips.

► Following the Revolution, Paris regained its position as the centre of power in France, which it had lost when Louis XIV moved his court to Versailles. Under Napoleon, the transformation of the capital into a modern city began, with the construction of new boulevards, river embankments and bridges, such as the Pont de la Concorde.

city in the world, was already approaching 1 million. Meanwhile, the USA was developing a strong industrial base in the expanding centres of New York, Chicago and Philadelphia. Within barely a century, its might would outstrip the Old World.

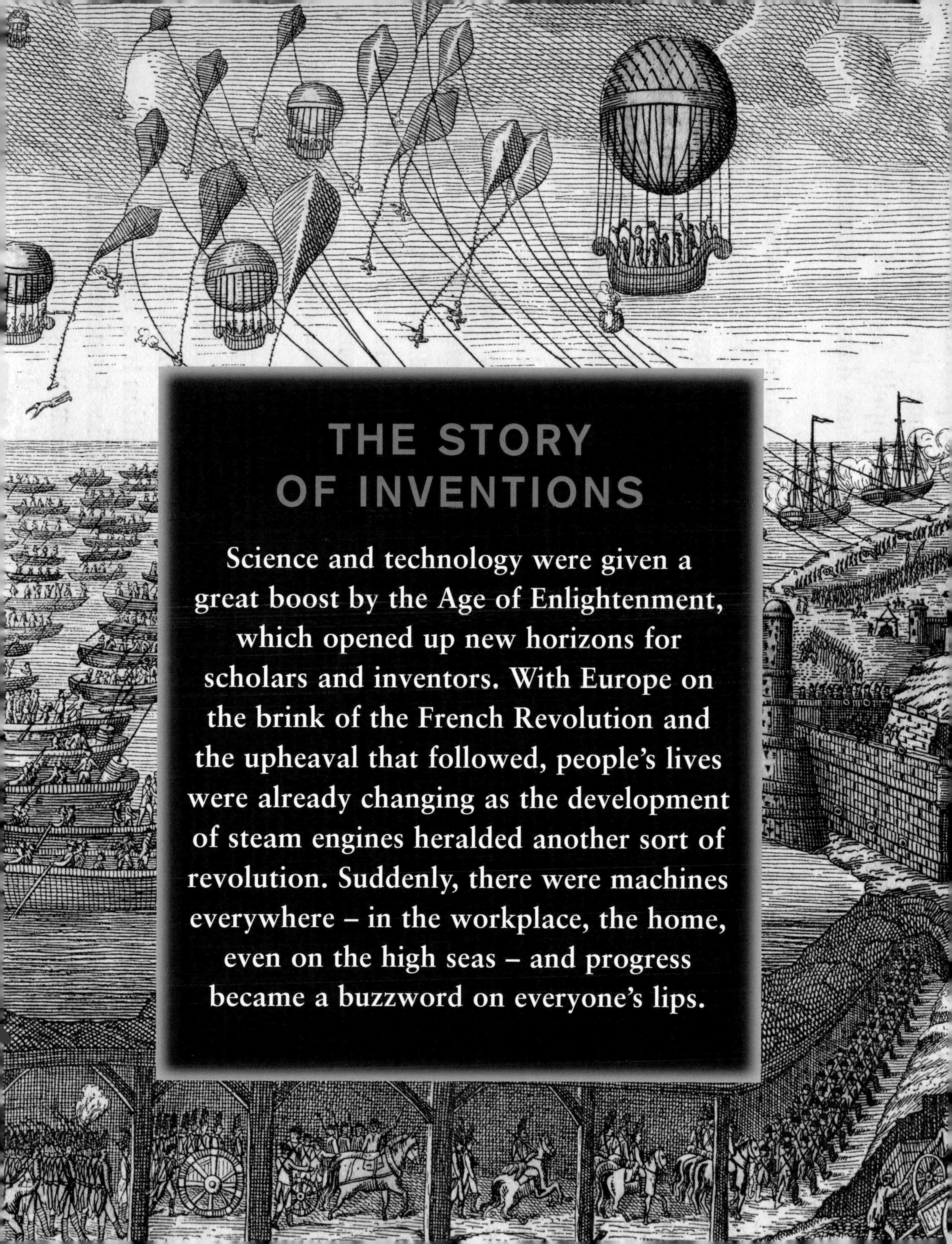

THE STORY OF INVENTIONS

Science and technology were given a great boost by the Age of Enlightenment, which opened up new horizons for scholars and inventors. With Europe on the brink of the French Revolution and the upheaval that followed, people's lives were already changing as the development of steam engines heralded another sort of revolution. Suddenly, there were machines everywhere – in the workplace, the home, even on the high seas – and progress became a buzzword on everyone's lips.

THE LIGHTNING CONDUCTOR – 1753

From bolts to volts

Lightning became the subject of scientific study in the 18th century through groundbreaking work by a handful of brave pioneers, notably Benjamin Franklin. During a thunderstorm in Philadelphia, this respected American statesman and scientist flew a kite fitted with a metal conductor close to a storm cloud and succeeded in producing an electrical discharge. Franklin's bold experiment proved that lightning was electricity and prompted him to invent the lightning conductor.

Steeples and towers
Some of the models and instruments used by Benjamin Franklin in his experiments with atmospheric electricity (above).

Since the very earliest times, people have been terrified by lightning. The Greeks thought that thunder and lightning were produced by the Cyclopes, a race of fearsome one-eyed giants who forged thunderbolts for Zeus to use as weapons. The Romans regarded lightning as a clear sign of the capriciousness of their gods. And whenever dark clouds filled the sky in the Middle Ages, Christians would call upon various saints, including St Donatus and St Barbara, for divine protection.

Scholars began to distance themselves from such superstitions in the 17th century, but their explanations of the phenomenon were no less wide of the mark. In his 1637 essay on meteors, for example, René Descartes clung to an idea first advanced by the Roman philosopher Seneca, who proposed that thunder arose when heavy, high clouds descended and collided with a layer of lower clouds. Compressed by this sudden, violent movement, the air between the two cloud layers not only generated the familiar rumbling sound, but also gave off heat, which was the source of lightning flashes.

AWESOME POWER

When a lightning strike hits the ground, it discharges a billion volts within a fraction of a second.

Franklin's hunch

This theory was exploded by the self-taught American scientist and scholar Benjamin Franklin (1706–90), who began to show an interest in electricity in 1746. Franklin's achievements in this field were extraordinary. He identified positive and negative charges – one of the cornerstones of physics – and, following his realisation that electricity was fluid like a liquid, he was the first to suggest the idea of an electric current. He maintained that a lightning bolt was nothing more than an extremely powerful electrical spark, and that the electricity came from the skies.

Franklin's observation that metal points concentrated electrical charges led him to devise an experiment. Initially, he did not put his idea to the test in person, but he did put pen to paper and described the principle in a letter written on 29 July, 1750, to his friend Peter Collinson, a physicist and Fellow of the Royal Society in London: 'On the top of some high tower or steeple, place a kind of sentry box big enough to contain a man and an electrical stand. From the middle of the stand let an iron rod rise and pass bending out of the door, and then upright 20 or 30 feet, pointed very sharp at the end.'

Collinson immediately recognised the importance of the work that Franklin was pursuing. The following year he published Franklin's letters in a volume entitled *Experiments and Observations on Electricity, made at Philadelphia in America*. In 1752 the naturalist and Director of the Royal Botanical Gardens in Paris, Georges-Louis Leclerc, Comte de Buffon, had the book translated into French by one of his assistants,

MARTYR TO SCIENCE

Several scientists in the 18th century attempted to 'capture fire from the sky', but unfortunately some of them failed to take adequate precautions. The German physicist Georg Wilhelm Richmann was experimenting with lightning in St Petersburg, where he had installed a metal rod running from the roof of his house to his study. In August 1753 he was struck dead by a bolt of lightning that travelled down the rod. Mikhail Sokolaw, engraver to the Academy of St Petersburg, witnessed the fatal strike. He was blown off his feet and knocked unconscious, but survived.

LA SCIENCE POPULAIRE
JOURNAL HEBDOMADAIRE ILLUSTRÉ
Rédacteur en chef : ADOLPHE BITARD. BUREAUX : rue Montmartre, 125.
N° 5. — Prix : 15 centimes.
Abonnements. — Paris, un an, 8 fr. ; six mois, 4 fr. — Départements, un an, 10 fr. ; six mois, 5 fr.
SOMMAIRE

Thomas-François Dalibard, then asked him to attempt the experiment suggested by Franklin. Dalibard gave instructions to a joiner by the name of Coiffier, who erected a 13-metre-high iron rod, resting on an insulating support and kept in place by silk cords, at Marly-la-Ville outside Paris. On 10 May, 1752, while a thunderstorm raged above the town, Coiffier succeeded in drawing sparks from the

Proving Franklin right *On 10 May, 1752, at Marly-la-Ville near Paris, an experiment conceived by Benjamin Franklin was put into effect by French scientist Thomas-François Dalibard, proving that bolts of lightning are charged with electricity. This unsigned engraving of the event (right) was made in the 19th century.*

Struck dead *The French magazine* La Science Populaire *(left) devoted the front page of an issue in March 1880 to the death of Georg Richmann, killed while studying lightning.*

A FLASH OF GENIUS

In the 1st century BC, in his poem *De Rerum Natura* ('On the Nature of Things'), the Roman philosopher Lucretius was well ahead of his time in being one of the first people to offer an explanation for the cause of thunder that did not attribute it to divine intervention. Facetiously, he wondered why Jupiter should always choose to appear when thunderclouds were looming and never when it was sunny. In his poem Lucretius referred instead to the 'primordial atoms of the thunderbolt', a physical explanation more akin to modern scientific theory, which places the origin of electrical storms in the friction between particles in a cloud.

Natural spectacle *In the 18th century, various implausible theories were put forward to explain the causes of lightning. In fact, a lightning bolt is simply a discharge releasing the build-up of electrical charge in a cloud.*

thundercloud – and survived to tell the tale. Franklin's theory that thunderclouds were charged with electricity was thereby proven. Three days later, Dalibard concluded his report on the matter to the French Academy of Sciences in the following terms: 'Without a doubt, the substance of thunderstorms is the same as that of electricity. Mr Franklin's theory is no longer merely a conjecture.'

Running to earth

On the other side of the Atlantic, Franklin knew nothing about Dalibard's demonstration at Marly and was busy devising another experiment to confirm his idea. It famously involved a kite.

Franklin did not leave an account of the experiment himself, but the version that has come down through history recounts that he tied a large silk handkerchief over two sticks lashed together to make a kite, then mounted a thin iron spike on the top. He tied the kite to a line of moistened hemp twine, which would conduct electricity but not too efficiently, and near the end of the twine he fastened a metal key, which would become electrified in the event of a lightning strike. He tied the kite string to an insulating silk ribbon. Early in June 1752, as a thunderstorm rumbled around his home town of Philadelphia, Franklin raised the kite into the storm, regardless of any danger to himself. As he saw the key receive an electrical charge, he had his proof of the link between lightning and electricity.

Flying a kite
According to popular legend, as reflected in this illustration (right), Franklin's young son accompanied him during his kite experiment to prove that lightning was electricity. There is some debate as to how the experiment was actually conducted. His son would have been around 21 years old at the time, not a boy, and some accounts say that Franklin flew the kite from an upper window. Some even claim that the experiment as shown would have been fatal.

News of Franklin's successful experiment quickly spread, and a craze arose throughout Europe for flying kites in electrical storms. Franklin, ever the pragmatist, put his research into practice, seeking ways of protecting people and property against lightning strikes. Why not, he reasoned, fit insulated metal rods to tall buildings, sticking up from the very top and running all the way down to the ground? The electrical discharge caused by a lightning flash would strike the rod, rather than the building, and travel down the metal to flow harmlessly into the ground – or to 'earth' in a phenomenon that became known as earthing.

MAN OF MANY PARTS

Benjamin Franklin, shown here in a three-dimensional model (right), was born in Boston in January 1706, one of 17 children of Josiah Franklin, a soap and candle manufacturer. An avid reader and tireless worker, the self-taught Benjamin produced a number of inventions besides the lightning rod, including a convection stove to distribute heat throughout a house, a musical instrument known as the glass armonica and bifocal spectacles. In 1776 he was co-author, with Thomas Jefferson and John Adams, of the US Declaration of Independence, and he also had a part in drafting the Federal Constitution of 1787. In 1776, he travelled to Paris to persuade the French king, Louis XVI, to support the American colonists' attempt to break free from Britain. His efforts bore fruit in 1778 in a trade and friendship treaty between the two nations, followed by French military intervention from 1780 in the American War of Independence. The French economist and statesman Anne-Robert-Jacques Turgot proposed this motto for an inscription on a bust of the great man: *Eripuit fulmen coelo sceptrumque tyrannis* – 'He snatched the lightning from the sky and the sceptre from tyrants'. Franklin died in Philadelphia in April 1790.

LIGHTING UP HIS HOME

In the early 1750s, Franklin set up a lightning rod on top of his house, connected to a bell; a second bell was connected to a grounded wire. He surmised that, in an electrical storm, the bells would ring when hit by a brass ball suspended between them on a silk thread. What he actually observed was quite different: 'I was one night awakened by loud cracks on the staircase. Starting up and opening the door, I perceived that the brass ball, instead of vibrating as usual between the bells was repelled and kept at a distance from both; while the fire passed, sometimes in very large, quick cracks from bell to bell, and sometimes in a continued, dense, white stream, seemingly as large as my finger, whereby the whole staircase was inlightened [sic] with sunshine, so that one might see to pick up a pin.'

Man of the moment
An 18th-century engraving shows Benjamin Franklin carrying an example of a 'lightning-rod umbrella'.

An instant hit

Franklin first described the principle of his lightning conductor in 1753 in *Poor Richard's Almanack*, an annual publication that he himself had founded and edited since 1732. He then went on to test out his device on his own house in Philadelphia.

Franklin's 'single-spike lightning rod' was an instant hit in the United States: within less than 10 years, 10,000 had been installed on the roofs of dwellings and the towers and spires of churches. Thereafter, beginning in Bavaria, it found favour in Europe. In Venice, the buildings of the Arsenale were fitted with lightning rods to protect the fleet from fire. In 1770, St Paul's Cathedral in London became the first building in Britain to have the safeguard of a lightning conductor.

In France, meanwhile, it was argued by some scientists that rods attracted lightning to buildings, rather than protecting them, so it was not until 1776 that France got its first lightning conductor, on the Church of St Philibert in Dijon. Yet by the century's end the country had caught lightning-rod fever. Fashionable people on the streets of Paris even sported umbrellas and hats equipped with a metal spike earthed to the ground by a thin metal chain.

Drawing fire
This dramatic image of the Eiffel Tower (now in the Musée d'Orsay in Paris) was taken in 1902 by French photographer Gabriel Loppé. It shows the 300-metre cast iron structure being struck by lightning. On completion in 1889, the metal tower was the tallest building in Paris and it acted as a giant lightning conductor for the entire city.

THE *ENCYCLOPÉDIE* – 1751 TO 1772

A key text of the Enlightenment

The *Encyclopédie*, or 'Systematic Dictionary of the Sciences, Arts and Trades', compiled under the leadership of Diderot and d'Alembert, embodied a rational approach to knowledge that was in stark contrast to traditional religious teaching. This work paved the way for new areas of scholarship and enlightened modes of thought that led directly to the French Revolution and continued to resonate for a long time thereafter.

Leaders of enlightenment
Portraits of D'Alembert (above) and Diderot (top) as they appeared in the first volume of their encyclopedia.

Originally, the *Encyclopédie* was conceived as a French translation of the bestselling three-volume Chambers' *Cyclopedia*, published in London in 1728. Parisian publisher André Le Breton applied for a royal warrant in 1747 to produce an expanded translation of this work, entrusting editorial direction to the Abbé Gua de Malves, a member of the French Academy of Sciences. But no sooner had the project begun than it was shelved in favour of a far more ambitious undertaking, funded by subscribers. In 1750 the controversial novelist and philosopher Denis Diderot, who only the previous year had been imprisoned for an essay that questioned the existence of God, replaced Malves as general editor. Inspired by the 17th-century English philosopher and scientist Francis Bacon, Diderot drew up a new prospectus for the work. The *Encyclopédie* would, he announced, set forth a 'taxonomy of human knowledge' which placed Man rather than God at the centre of the universe.

IN FIGURES

It took more than 1,000 collaborators 15 years to complete the core volumes of the *Encyclopédie*. The 75,000 entries covered 18,000 pages with 20.8 million words and 2,885 illustrations.

Instant infamy

The first volume of the *Encyclopédie* proper was published in 1751, with a 'preliminary discourse' by the mathematician Jean le Rond d'Alembert that reads like a manifesto for the Age of Reason. The second volume appeared the following year. Certain articles were highly controversial. In the entry on 'Certainty', for example, the assertion that this must be based on empirical proof flatly contradicted Christian belief. The Jesuits launched such a sustained and effective campaign against the work that the Royal Council issued a ruling banning the sale of the first two volumes. Others came out in support of the *Encyclopédie*'s compilers, notably Chrétien de Malesherbes, the curator of the Royal Library, who was responsible for regulating and censoring the book trade. Publication duly resumed in 1753, though henceforth d'Alembert's contribution, perhaps wisely, was restricted to editing entries on mathematics.

On 5 January, 1757, a failed attempt on the life of Louis XV at Versailles gave the religious lobby a pretext to attack the *Encyclopédie* for supposedly fostering a climate of anarchy.

FORERUNNERS OF THE *ENCYCLOPÉDIE*

In the 1st century BC, Roman scholar Marcus Terentius Varro became the first author of an encyclopedia-like tome, the *Antiquitates rerum humanarum et divinarum* (now lost). Later works of this type included Pliny the Elder's *Natural History* (AD 77), the *Dictionarum universale* of Bishop Solomon III of Constance (9th century) and Vincent of Beauvais' *Speculum majus* (13th century). Major reference works in later centuries include the *Dictionnaire historique et critique* (1696) by the French Huguenot scholar Pierre Bayle and Ephraim Chambers' *Cyclopedia* (1728).

Powerful adversaries
The Encyclopédie *united Jesuits, Jansenists and the Pope against it. They managed to delay individual volumes and to censor certain entries, but failed to halt the onward march of this great work (above). It eventually stretched to 35 volumes, published between 1751 and 1772.*

Table talk
Seated round the dinner table in this 18th-century engraving are leading figures of the French Enlightenment, including Voltaire, Diderot, the Abbé Maury and Condorcet.

They succeeded in getting the work's royal patronage revoked. Publication was halted once again after Volume VII, when the Pope placed the *Encyclopédie* on the Church's list of banned books. D'Alembert now relinquished his involvement in the project and several other editors also quit. Another intervention by Malesherbes managed to get the ban partially lifted, allowing the publication of 11 less controversial volumes of engravings illustrating the industrial arts to compensate

THE 'CACOUAC' TRIBE

In October 1757, an anonymous article appeared in the newspaper *Mercure de France*, entitled 'A Salutary Warning, or First Report on the Cacouac Tribe'. This nonsensical name referred to the compilers of the *Encyclopédie*. It was gleefully taken up by their enemies, who included the conservative historian Jacob-Nicolas Moreau and the journalist Élie Fréron, a sworn enemy of Voltaire. Enlightenment thinkers were lampooned as a tribe of savages 'wilder than the Caribs ... the only creatures in the natural world who do evil for its own sake'. Instead of rising to the bait, the encyclopedists playfully adopted the term as a badge of pride.

ILLUSTRATING KNOWLEDGE

The *Encyclopédie* is justly famous for its meticulous illustrations. Initially, the idea was to copy and incorporate as many existing drawings as possible, such as the anatomical sketches of Vesalius. Yet this only led opponents of the project to accuse its authors of plagiarism, so it was decided to commission brand new plates. Fully 900 of the total 2,885 copperplate engravings were the work of the brilliant Parisian illustrator Louis-Jacques Goussier. He went to the source to ensure the accuracy of his drawings, visiting forges, factories and mines. Diderot claimed that Goussier was responsible for 'all the great plates in our Encyclopedia', and immortalised him in the character of Gousse in his novel *Jacques le Fataliste* (1796).

Visual quality *A cutaway engraving of a mine displays the realism and detail achieved by Goussier through direct observation. Many regard him as the third encyclopedist, alongside Diderot and d'Alembert.*

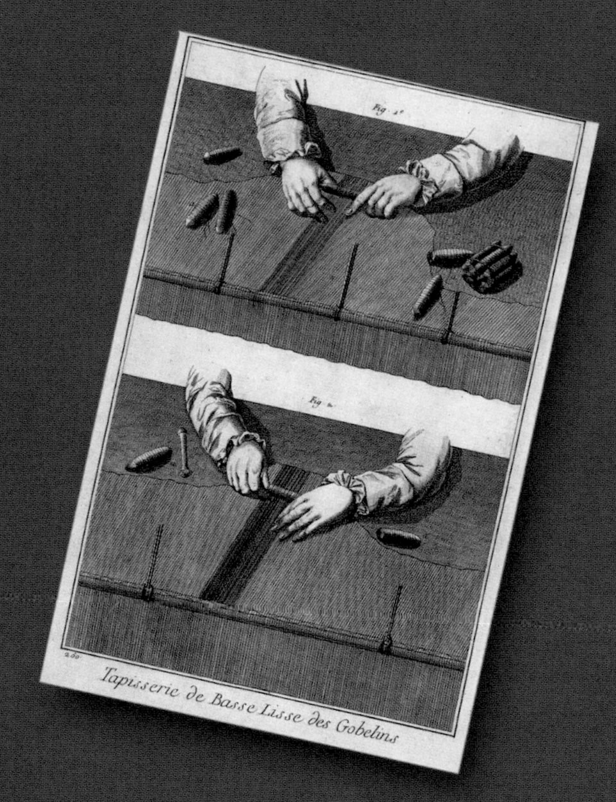

subscribers. Meanwhile, editing and printing of the text volumes continued in secret. The final 10 volumes of text were eventually published in 1765. In places, these bear the marks of self-censorship to preempt official sanctions: to Diderot's great displeasure, le Breton summarily dropped several articles.

An army of editors

In his prospectus, Diderot announced that a team of 55 editors had been assembled to work on the encyclopedia, including the cream of French literature and science. In fact, over the years the number of contributors swelled to 160 or more, while the total number of people involved in the project rose to more than 1,000. The editors were of variable quality, as Diderot later conceded: 'Alongside some excellent editors, there were also some weak, mediocre or even downright bad ones. The end result, then, was something of a curate's egg, where hack's scribblings rubbed shoulders with some minor masterpieces'.

Illustrious names include Voltaire, the academics Marmontel and La Condamine, Turgot, Necker and Rousseau. Rousseau edited dozens of articles on music and also the entry on 'Political economy', but he withdrew from the project after his article on Geneva whipped up a storm of controversy. Less well-known editors – from the middle classes, the aristocracy and the Church (Catholic and Protestant) – were recruited for their specialist knowledge of the law, medicine, mathematics, chemistry, agriculture or religious affairs.

The editors were not all French. A smattering of experts came from other European countries such as Switzerland,

A PROLIFIC CONTRIBUTOR

During the encyclopedia's eight-year ban, Huguenot scholar Louis de Jaucourt edited no fewer than 17,000 articles, almost half of all the entries in the second series. A highly cultured man, de Jaucourt was a member of the Royal Society in London, and was versed in theology, the sciences and medicine. He wrote a life of Leibniz, and drafted a major encyclopedic dictionary of medicine. This latter task took him 20 years, but was ultimately in vain – the manuscript was lost at sea. De Jaucourt was the author not only of many scientific treatises but also influential essays on such diverse topics as 'Equality', 'Monarchy', 'The Motherland' and 'The Treatment of Negroes'. In this last article, he condemned the purchase of negroes for the purpose of reducing them to slavery. The slave trade, he argued, 'flies in the face of all the precepts of Christianity, morality, natural law and human rights'.

Flight of fancy
One of the less realistic engravings in the encyclopedia is this whimsical gardener from the royal palace at Versailles.

Workmen's tools and techniques
Many copperplate engravings in the Encyclopédie *were designed to show the tools and working methods of artisans in various industries. The plate above, for example, shows the tools used by a glazier. The plate above left shows how a worker weaves a shuttle between the threads on a low-warp loom at the renowned Gobelins tapestry manufactory in Paris.*

Man of Letters
A portrait of the historian Jules Michelet (1798–1874), who called the Encyclopédie *a 'benign conspiracy that spread to take in the entire world'.*

Weighty tomes
An antiquarian bookshop in Barcelona, Spain. Antiquarian books are divided into several different categories: incunabula *comprise the oldest books, dating from the very early days of printing and published before 1501; antiquarian works proper date from the 16th to 18th centuries; Romantic works are from the period 1820–50, while books published between 1850 and 1914 are known as modern.*

Poland, Italy and Bavaria. One contributor was the banker and Enlightenment thinker Baron d'Holbach from the Rhineland Palatinate. He had written a number of anticlerical pamphlets under various pseudonyms before being invited to contribute articles on mineralogy, metallurgy, chemistry and politics to the encyclopedia. His friend Claude Adrien Helvétius, head of the French customs and excise, provided financial backing for the enterprise. As French historian Jules Michelet later remarked, the *Encyclopédie* was 'much more than just a book; it was a sect comprising all the finest minds in Europe'.

Revolutionary ideas

Yet the real novelty of the *Encyclopédie* lay in its approach to learning. The system adopted by Diderot and d'Alembert was designed to establish the affinities between various disciplines, broadly divided into three categories: history, the arts and the sciences. The arts comprised both the seven 'liberal arts' of Classical learning (grammar, rhetoric, logic, arithmetic, geometry, music and astronomy) and seven 'mechanical arts' (weaving, agriculture, architecture, warfare, commerce, cooking and metalurgy). Pride of place was given to philosophy, of which theology was considered merely an adjunct. As d'Alembert wrote: 'Knowledge derives not from Rome or from the Bible, but from the intellect'. Diderot was equally clear and ambitious about the work's agenda: 'In time, this work will surely spark an intellectual revolution, which I sincerely hope will bring about the downfall of tyrants, oppressors, fanatics, and bigots.'

While it did not explicitly urge the common people to take up arms nor espouse Rousseau's ideal of equality, the *Encyclopédie* did broadly promote the idea of liberty. Both in the subject matter of such entries as 'Political authority', 'Natural law', 'Oppressor' or 'The People', and in their own struggle to bring their project to a successful conclusion, the encyclopedists epitomised freedom of thought in the face of repression. Their work prefigured the French Revolution and laid the intellectual groundwork for the Industrial Revolution.

A world of learning

Despite the high cost, the first edition of the *Encyclopédie* comprised 4,250 copies, a large

VOLUME BY VOLUME

Between June 1751 and December 1765, the combined efforts of four Parisian publishing houses – le Breton, Briasson, David and Durand – saw the publication of 17 text volumes of the *Encyclopédie*. Due to the extended ban on the work, publication took place in two phases. The names of Diderot and d'Alembert appear on the title pages of the first series (volumes I to VII), but are missing from the second series (VIII to XVII). Aside from the odd frontispiece, marginal illustration or ornate capital, the entries in these text volumes are not illustrated. In 1762 the first of 11 volumes of plates appeared. This series, which ran until 1772, was entitled *A collection of plates on the sciences and the liberal and mechanical arts, fully captioned and annotated.* A supplement entitled *New Dictionary: A supplement to the dictionary of the sciences, arts, and crafts* was published in Paris from 1776 to 1780 by the firm of Panckoucke, Stoupe and Brunet, and in Amsterdam by Rey. It comprised four volumes, plus an extra volume of engravings and two volumes of index, making 35 volumes in total.

print run for the time. By 1768, the volumes acquired by the first subscribers were reselling at up to 15 times their original price. Also, several editions were printed quite independently of Diderot, with the result that there were some 24,000 copies of the work in circulation by the time of the French Revolution in 1789.

The *Encyclopédie* also inspired similar projects elsewhere: the *Encyclopaedia Britannica,* which was originally published in three volumes in Edinburgh in 1768–71, had grown into a 30-volume set by the time the 15th edition was printed in Chicago over 200 years later. In France, Larousse launched the *Grand Dictionnaire universel du XIX siècle* in 1863, followed by a whole range of encyclopedias in the 20th century. Other notable general encyclopedias from the 19th and 20th centuries include the *Encyclopedia Americana*, the German *Brockhaus*, the *Dizionario enciclopedico italiano* and the Russian *Bolchaïa Sovietskaïa Entsiklopedia.*

From the early 1990s onwards encyclopedias began to appear in digitised form on CD-ROMs. Nowadays, internet search engines direct people to sites such as the Encyclopaedia Universalis or Wikipedia. This latter site is a collaborative encyclopedia project, where anyone is free to contribute and alter the content of articles. As such, it comes with the disclaimer that it 'cannot guarantee the validity of the information found here'. A Wikipedia user, then, would be well advised to use those same critical faculties that were first promoted by Diderot and d'Alembert in their *Encyclopédie.*

Knowledge acquisition
The traditional method of learning is to visit a library, like these scholars in the Bibliothèque Nationale in Paris (above), but people are increasingly turning to Web sources such as Wikipedia for their information.

INTRIGUING CROSS-REFERENCES

Long before the Web came up with the concept of hyperlinks, the *Encyclopédie* introduced an ingenious system of cross-references, designed not just to tie the work's alphabetically organised articles into a broader framework, but also to inspire readers to explore interesting related topics and to bamboozle the censors. For example, the relatively neutral entry on 'The Franciscans' cross-referred to 'The Capuchins', in which all religious orders were lampooned. Diderot intended the cross-references to be of four distinct types, focusing on:

- **words – pointing the reader to a definition in another entry;**
- **things – corroborating or refuting an idea contained in one article by reference to another;**
- **satire – epigrammatic references that served to undermine received ideas (the entry on 'Cannibals' ends with a cross reference to 'Eucharist, Communion', i.e. the 'body of Christ');**
- **inspiration – speculative links aimed at creating wholly new associations of ideas and insights; Diderot defined these 'brainstorming' references as 'the extravagant conjectures of a genius'.**

PUBLIC MUSEUMS – 1759

Culture for all

The first public museums began to appear in Europe in the mid-18th century. The first national public museum was the British Museum, which opened its doors in 1759. Public museums really took off from the 1780s onwards, borne on a tide of enthusiasm for conservation and the desire of nation-states to promote their cultural and artistic heritage.

Viewing antiquities *Visitors admire the Sculpture Room in the British Museum in the early 19th century. The room is also known as the Elgin Room, after Thomas Bruce, Lord Elgin, the British Ambassador to Constantinople in the late 18th century. In 1799, Lord Elgin famously bought large sections of the decaying marble frieze around the Parthenon in Athens and had them shipped back to Britain where they were conserved. More recently, the Greek government has requested the return of the Elgin Marbles and their ownership remains a controversial issue.*

Mighty Roman emperor *A bust of Constantine I, the Great, which was first displayed in the Forum in Rome in the 4th century AD. It is now on show in the Capitoline Museum.*

The first public museums were born out of royal collections and 'cabinets of curiosities' belonging to private collectors. On his death in 1753, Sir Hans Sloane, physician to King George II and a keen amateur naturalist, left behind his personal treasure trove – a vast and eclectic collection of stuffed animals and birds, herbaria, engravings, medallions and other items brought back by travellers from far-off lands or acquired through the wholesale purchase of other collectors' cabinets of curiosities. For the tidy sum of £20,000, the state bought these treasures from Sloane's heirs and used them to form the core of the newly founded British Museum, which opened its doors to the public in 1759.

WHAT'S IN A NAME?

The *New World of Words*, an English dictionary published in 1706, defines 'museum' (from the Greek *museion*) as 'a public place for the recreation of learned men'. This definition recalls the original purpose of the museum in antiquity. At the Museion in Alexandria, founded in 300 BC, philosophers met to study and pay homage to the Muses, the goddess-patrons of the arts and sciences to whom the institution was dedicated. In Diderot and d'Alembert's *Encyclopédie*, the definition had become a 'place containing objects with a direct relation to the Arts and the Muses'.

A slow process

Since the earliest times, princes had been in the habit of collecting portraits and other exquisite *objets d'art*. In the Renaissance, it became the fashion for rich collectors to amass objects from every conceivable field of learning, from archaeology to zoology. In a treatise of 1565, the Flemish scholar Samuel von Quiccheberg, a doctor in the service of Albert V of Bavaria, neatly summed up the purpose of an ideal museum as being to bring together objects from the entire known world and display them according to a logical organisational system.

These prototypic museums – known as 'cabinets of curiosities' in English, *studioli* in Italian and *Wunderkammer* in German – were not merely designed to display the wealth and taste of their owners. They also had an educational agenda. And the more opulent the collection, the greater the chance that it would outlive its creator intact.

In 1471, Pope Sixtus IV bequeathed his collection of ancient sculptures to the city of Rome, so laying the foundation of what would become the Capitoline Museum. In 1671, the city and university of Basel in Switzerland acquired the Amerbach Collection, a renowned cabinet of curiosities. Care was taken to protect the exhibits from overzealous visitors – in the early days of museums, people's propensity to handle objects was a common problem. In 1710, a German scholar visiting the Ashmolean Museum (founded in 1683 under the auspices of the University of Oxford) was reportedly outraged by this barbarous habit.

Museums went from strength to strength. In 1719, Tsar Peter the Great ordered that a room of rare items he had collected be opened to his subjects in order that they should have the opportunity of improving themselves. In 1750, Louis XV agreed to let the public view the royal collections at the Luxembourg Palace twice a week.

Conservation and display

Soon after their inception, museums began to specialise. At the Uffizi Palace in Florence, botanical and zoological specimens and scientific instruments, hitherto displayed alongside the works of art, were hived off into a separate area devoted to physics and natural history. The collection of paintings was reorganised to give prominence to works by Tuscan artists, and a purpose-built gallery to house them was opened in the late 1780s.

Before long, museums were springing up all over Europe: the Teyler Science Museum opened in the Dutch city of Haarlem in 1778, while 1792 saw the founding of the Prince-elector's gallery in Düsseldorf and the Belvedere in Vienna – open to all, 'provided that they have clean shoes'. The following year, the Louvre Museum was established in Paris, with the express intent of bringing under one roof all the treasures seized from the clergy and the crown during the Revolution, to preserve them as examples of the nation's cultural heritage. Yet the Golden Age of museums in Europe's imperial capitals was still to come, with the acquisition in the 19th century of artefacts from the great civilisations of antiquity, such as ancient Greece and Egypt.

Perusing portraits
A view of the Portrait Room at the Vienna Belvedere. Designed by Johann Lukas von Hildebrandt, this former summer palace of Prince Eugen has been a public museum since 1792.

Tsarist treasure
St Petersburg's Hermitage Museum (below) has one of the world's largest collections – some 60,000 artefacts, in 1,000 rooms.

FAKING IT

In 1842, a hideous mummified creature, apparently half-human, half-fish, caused a sensation when shown in New York. The so-called 'Feejee mermaid' was later unmasked as a fake – the head of a monkey grafted onto a fish tail, a publicity stunt by the showman P T Barnum. Perhaps the most infamous hoax was Piltdown Man, allegedly the remains of an early hominid found in southern England in 1912. The skull turned out to be that of a modern human, with the jaw of an orang-utan.

Roller skates c1760

Early exponent
A roller-skater in Holland shows how its done in August 1790. In the early days, skates were fitted with just two wheels in-line and had simple spur brakes, like those on ice skates.

The millions of devotees of skateboards and in-line skating worldwide owe a debt of gratitude to Jean-Joseph Merlin, clockmaker and musical instrument manufacturer. Merlin hailed from Belgium but ran a shop in London, where he first had the thought of adapting the sport of ice-skating to dry land by mounting two small metal wheels on the base of a wooden board. The idea was born in the 1760s, but failed to catch on due to one crucial omission – any means for the skater to stop safely. This design flaw earned Merlin the dubious distinction of becoming the first victim of a roller-skating accident.

Merlin's crash was spectacular and might have sounded the death knell for roller skates, had it not been for a French engineer by the name of Petibled (his first name is not known). In 1819, Petibled lodged a patent for a three-wheeled in-line skate, with a screw fitted to the heel as a brake. Four years later, a fruiterer named Robert John Tyers of Piccadilly, London, perfected an in-line skate with five wheels, which he christened the *Volito* (from the Latin *volare*, meaning 'to fly' or 'to flutter'). Tyers described it as an 'apparatus to be attached to boots ... for the purpose of travelling or pleasure'.

MR MERLIN'S MISFORTUNE

Jean-Joseph Merlin came to grief while showing off his new invention – and attempting to play the violin at the same time – at a masquerade thrown by a Mrs Cornely, a fashionable London society hostess, at her assembly rooms in Carlisle House in Soho Square. A contemporary observer recorded what happened: 'Not having provided the means of retarding his velocity, or commanding its direction, he impelled himself against a mirror of more than five hundred pounds' value, dashed it to atoms, broke his instrument to pieces, and wounded himself most severely.'

Skating mania

Forty years were to elapse before the first pair of roller skates equipped with rubber cushion shock absorbers appeared. Their inventor, James L Plimpton of New York, is regarded by many as the father of the modern roller skate. Plimpton's 'quad skates' had four small boxwood wheels, mounted front and back in pairs on movable axles, which allowed skaters to turn without lifting their feet off the ground. Far more stable and manoeuvrable than earlier models, quad skates became an instant hit in America and remained the main form of roller skate until the reinvention of the in-line skate more than a century later.

Meanwhile, in 1852, the English inventor Joseph Gidman patented skates whose wheels had ball-bearing races, which made them faster and more responsive. Yet it was more than 30 years before this type of skate came into its own, when Chicago entrepreneur Levant Martin Richardson introduced a steel pin ball-bearing and began mass-producing skates. Roller skating now became a craze throughout the United States, as thousands of amateur skaters enjoyed the thrills and spills of the roller-skating rink. Rollermania spread to Europe via Britain.

Improved technology
These four-wheeled wooden skates of 1880 were fitted over the skater's ordinary shoes by means of straps and buckles.

a serious competitive sport, with the creation of official roller-skating federations.

Return of the in-line skate

The next major innovations were not until the 1970s, when polyurethane wheels came on the market. These were quieter than the earlier wooden or steel wheels. They also had greater manoeuvrability which made them more suitable for skating around pedestrians on city streets. Indeed, this virtually indestructible plastic gave roller-skating a new lease of life, paving the way for the hugely popular in-line Rollerblade, which was developed in 1979–83.

Grace and speed
While 19th-century adult skaters – like this couple from an advertisement (left) – took turns round the rink at a sedate pace, today's young skaters prefer things more fast and furious on state-of-the-art Rollerblades, with crash helmets, shin and elbow guards for protection (below).

In 1857 large public skating rinks opened in London at the Floral Hall in Covent Garden and in the Strand. The first roller-skating arena in Paris opened on the roundabout on the Champs-Elysées in 1875, with music provided by a 60-piece orchestra. Back in America, Madison Square Garden in New York became a skating rink in 1908.

For a long time, roller skating remained a genteel, middle class pastime for grown-ups, but gradually it began both to move down the social scale and to appeal to the younger generations. At the same time, it developed from being a purely recreational activity into

SKATES IN THE THEATRE

On 27 March, 1984, the rock musical *Starlight Express*, written by Andrew Lloyd Webber with lyrics by Richard Stilgoe and choreography by Arlene Phillips, opened to acclaim at London's Apollo Victoria Theatre. All the actors were on roller skates, playing a child's railway set that comes to life, and the theatre had been adapted to include a race track. The show was a sensation and ran for 7,461 performances.

Sandwiches 1762

Origin of the hamburger
A poster advertising the Hamburg to America shipping line (right), designed by Theodor Etbauer in 1937.

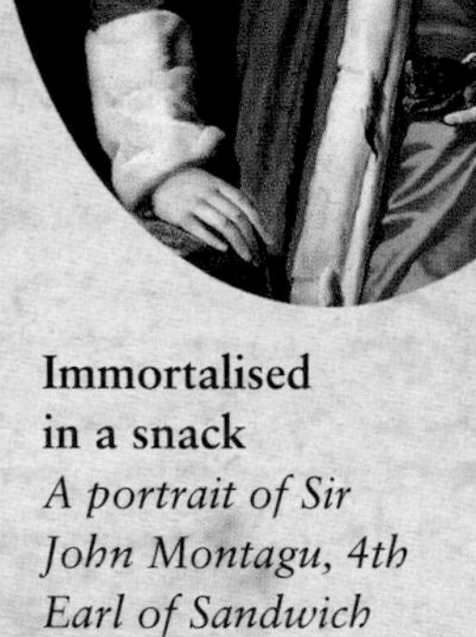

Immortalised in a snack
A portrait of Sir John Montagu, 4th Earl of Sandwich (above), by Joseph Highmore (1753).

Infinite variety
A club sandwich (right), one of countless variations on a theme.

John Montagu, 4th Earl of Sandwich (1718–92), was an avid habitué of London's casinos. He was always loath to interrupt an evening of cribbage or other card games with the tedious business of dinner. His cook came up with an ideal solution: placing cold meat or cheese between two slices of buttered bread, he brought his master and guests a substantial snack that they could eat with their hands while continuing their gaming.

Soon, people were ordering 'the same as Sandwich', and in no time this culinary brainwave spread beyond Britain's shores. American comic Woody Allen would later hail the sandwich as having 'freed mankind from the hot meal'.

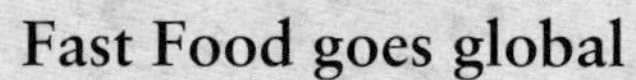

Fast Food goes global

Another famous form of fast food, and a relative of the sandwich, was invented by German immigrants to America in the 19th century. Many of them crossed the Atlantic on ships of the Hamburg-Amerika Line. A favourite meal for passengers on board these liners was a patty of minced beef served with salad, sliced tomatoes, onions (raw or fried), gherkins (dill pickles) sliced in thick rounds and a spicy sauce. The German émigrés adapted this 'Hamburg steak' to suit the taste of their new homeland and the hamburger was born, customarily served in a split sesame bun.

Anti-German sentiment in America during the First World War saw the hamburger's name changed briefly to 'Salisbury steak', but the rise of this quick, handy hot snack was unstoppable. From the mid-20th century, American hamburger franchises established fast-food restaurant chains worldwide.

MEDITERRANEAN MAYO

Mayonnaise, which first began to appear on dining tables in the late 18th century, takes its name from Port Mahon, capital of the Balearic island of Minorca. After retaking Minorca from the British in June 1756, Louis François Armand du Plessis, duc de Richelieu (great-nephew of Cardinal Richelieu), is said to have tasted the egg-based sauce in a tavern in the port. Delighted, he brought it to France as 'sauce de Mahon' or 'sauce à la Mahonnaise'.

THE PERFECT SANDWICH

Geoff Nute, a senior research fellow of Bristol University, claims to have found the formula for the perfect cheese salad sandwich:

$$W = [1 + ((bd)/6.5) - s + ((m-2c)/2) + ((v+p)/7t)]\ (100+l/100)$$

W equals the thickness of cheddar (in mm), b the thickness of bread, d is the dough flavour modifier, s thickness of butter, m the thickness of mayonnaise, c the creaminess modifier, v thickness of tomato, p the depth of pickle, t the tanginess and l the thickness of the lettuce layer.

Jigsaw puzzles *c*1760

The jigsaw was almost certainly invented in about 1760 by London cartographer and engraver John Spilsbury, who came up with the idea as an educational tool to make the geography of the British Isles a more appealing subject to children. Pilsbury mounted one of his maps onto a sheet of hardwood and cut around the borders of each county with a fine-bladed marquetry saw.

Phileas Fogg dissected
A French jigsaw puzzle of scenes from Jules Verne's classic adventure story, Around the World in Eighty Days *(1873).*

All kinds of themes

In 1787 another Englishman, the print-seller William Darton, produced a puzzle showing portraits of all the kings and queens of England from William the Conqueror to George III. Before long, all kinds of images were being reproduced on jigsaws, including mythological scenes, landscapes and historic buildings. The advent of plywood, which was much lighter and cheaper than hardwoods, revolutionised the manufacture of jigsaw puzzles at the start of the 20th century and made them affordable to all. The first jigsaws printed on cardboard – a far easier material to cut by machine, using a die-cutter – appeared in the 1920s. In the 1930s new puzzles were given away weekly as advertising premiums in the USA and became extremely popular.

The banjo *c*1764

The history of the banjo is tied up with the history of the United States. It was introduced to the New World by black slaves who had been transported from Africa to work on southern plantations. The original instrument was made from a goatskin stretched over a circular wooden frame, fitted with a long bamboo or wooden neck with four catgut strings.

Around 1830 the banjo was taken up by white musicians, who added a fifth drone string, shorter than the others and plucked with the thumb. They blacked themselves up to look like negroes and performed in so-called 'minstrel shows' in travelling circuses. This set off a huge boom in the instrument's popularity. In 1866, there were reckoned to be more than 10,000 banjo players in Boston alone. After the First World War, the banjo was adopted by Dixieland jazz and dance bands, and later became a firm favourite with Country and Western musicians.

ORIGINS OF THE BANJO

Earlier instruments associated with the banjo are the *banza* (Portuguese guitar) and *banjar* (a folk instrument from Antigua). Ultimately the term is thought to derive from *mbanza*, the word for the thumb piano used by the Kimbundu people of Angola. The first recorded use of the term 'banshow' by African-Americans for the modern instrument is in 1764.

Pick a tune *A banjo player photographed in 1902. The instrument has its origins in Africa, where its ancestors are still played today, such as the* ngoni *in the Gambia, Mali and Côte d'Ivoire.*

PESTICIDES – 1763

Chemical warfare on pests

For centuries, people had used various rudimentary methods to protect their crops from infestation by insects and fungi and from being overrun by weeds. But as the 18th century progressed, the population boom created an urgent need to increase the acreage under cultivation and this demanded more effective pest control. Development of the first pesticides was just one of the advances made by the new science of chemistry.

More than blight
A plate from the Encyclopédie agricole Quillet, *a French reference work of 1930, showing just a few of the 200 or more diseases that can afflict the potato.*

In the mid-18th century Louis XV of France bought up all the land around Montreuil near Paris, with an eye to annexing it to his estate at Versailles. This rural area had a long tradition of market gardening, horticulture and silviculture – a happy hunting ground for all kinds of pests. In 1763, it became the site of the world's first documented attempt at pest control through the use of chemicals, when tobacco juice was sprayed on an aphid infestation. Yet the use of pesticides can be traced right back to antiquity.

Arsenic and bare hands

From as early as 2500 BC the Sumerians used sulphur to control insect pests, and in the 9th century BC the Greek poet Homer reported the use of sulphur for fumigation. Later, in the 1st century AD, the Roman naturalist Pliny the Elder recommended arsenic as a pesticide in his *Natural History* and also wrote that extract of fennel added to sulphur discouraged mosquitoes. The Chinese began to deploy compounds derived from arsenic for this purpose from the 16th century onwards.

The work of the Swedish botanist and taxonomist Carolus Linnaeus in cataloguing and naming many pests in the 1750s did much to help studies of pests and their effects. Even so, up to the 19th century most farmers had little more than their bare hands to use against the insects that ravaged their crops. Picking bugs off was the main method of protection.

Dealing with known pests was just one part of the problem. Hand in hand with the huge growth in population and rapid urbanisation that took place in the 19th century came a whole series of hitherto unknown diseases. The most devastating of these was the potato blight that ravaged Ireland and other parts of Europe from 1845 onwards. This was closely followed by a succession of diseases that decimated Europe's wine-growing regions: powdery mildew in 1852, phylloxera in 1865 and downy mildew in 1878. In 1917 the potato crop was in the firing line once more,

THE PLAGUES OF EGYPT

Pest damage was feared in Biblical times. According to the book of Exodus, when the Pharaoh refused to release the Israelites from slavery, God sent ten plagues down on Egypt, including an infestation of frogs and swarms of mosquitoes, flies and locusts.

A CHANCE DISCOVERY

Towards the end of the 19th century, the wine-growing region of Bordeaux was devastated by downy mildew and phylloxera. So it came as a surprise to botanist Alexis Millardet when, in 1882, he found healthy vines growing in the Médoc in southwest France. The manager of the estate mentioned in passing that they painted a dye on the vines nearest the roads to deter passers-by from eating the grapes. Millardet analysed the dye and found that it contained copper sulphate and hydrated lime. Recreated as 'Bordeaux mixture', the combination has been used since to combat a range of plant fungal diseases.

AN UMBRELLA TERM

The generic term 'pesticide' was coined in the early 20th century from English the 'pest' plus the suffix '-cide', from the Latin word *caedere*, meaning to slaughter or kill. It broadly denotes a wide range of chemicals used against crop parasites, be they microscopic, animal, or vegetable. It encompasses insecticides, herbicides and fungicides (used against insects, weeds and fungal infection respectively), nematicides (to kill plant parasitic nematode worms), rodenticides (against rodents) and various others.

when it was attacked by a plague of Colorado beetles. The human and economic cost of these infestations can be devastating. In Ireland, where potatoes were the staple diet of the people, the famines of 1845–49 are estimated to have claimed the lives of around 1 million people and forced the migration of a million more. Between 1875 and 1889, French wine production plummeted from 86.3 million hectolitres to just 23 million.

Modern pesticides

Farmers and wine growers in the late 19th century resorted in the first instance to plant-based pesticides such as pyrethrum (an insecticide extracted from the Dalmatian or Persian chysanthemum) or rotenone, which occurs naturally in the roots and stems of certain tropical plants. Before long, advances in the chemical industry brought new mineral compounds onto the market, based on salts of copper, arsenic or sulphur. In 1860 chemicals were deployed against the Colorado potato beetle in the form of copper acetoarsenite or 'Paris Green', a bluish-green substance first used to kill rats in Parisian sewers, hence its name. In 1878 the French chemist Ulysse Gayon and botanist Alexis Millardet developed the fungicide 'Bordeaux mixture', which protected vines from downy mildew (*Plasmopara viticola*) but was largely ineffective against Phylloxera.

The growth of organic chemistry (the study of carbon-based molecules) made it possible to manufacture active chemical compounds tailored to the specific needs of farmers: the era of synthetic pesticides had dawned. In 1932, the first selective herbicide, dinitro-ortho-cresol (DNOC) was patented in France. Seven years later came the world's first modern insecticide, dichlorodiphenyl-trichloroethane – DDT for short. DDT, which had first been synthesised in 1874, was embraced with enthusiasm around the world once its insecticidal properties were discovered, but it was banned in the 1970s after it was shown to be devastating the environment. It could remain active in soil for up to 30 years and had poisoned the food chain, wiping out many animal and insect species and leaving others on the very brink of extinction.

Dangerous chemical
The chemical DDT was used extensively during and after the Second World War as an insecticide against mosquitoes, which carry deadly diseases like malaria and typhus. Unfortunately, it was later found to be killing far more than the mosquitoes.

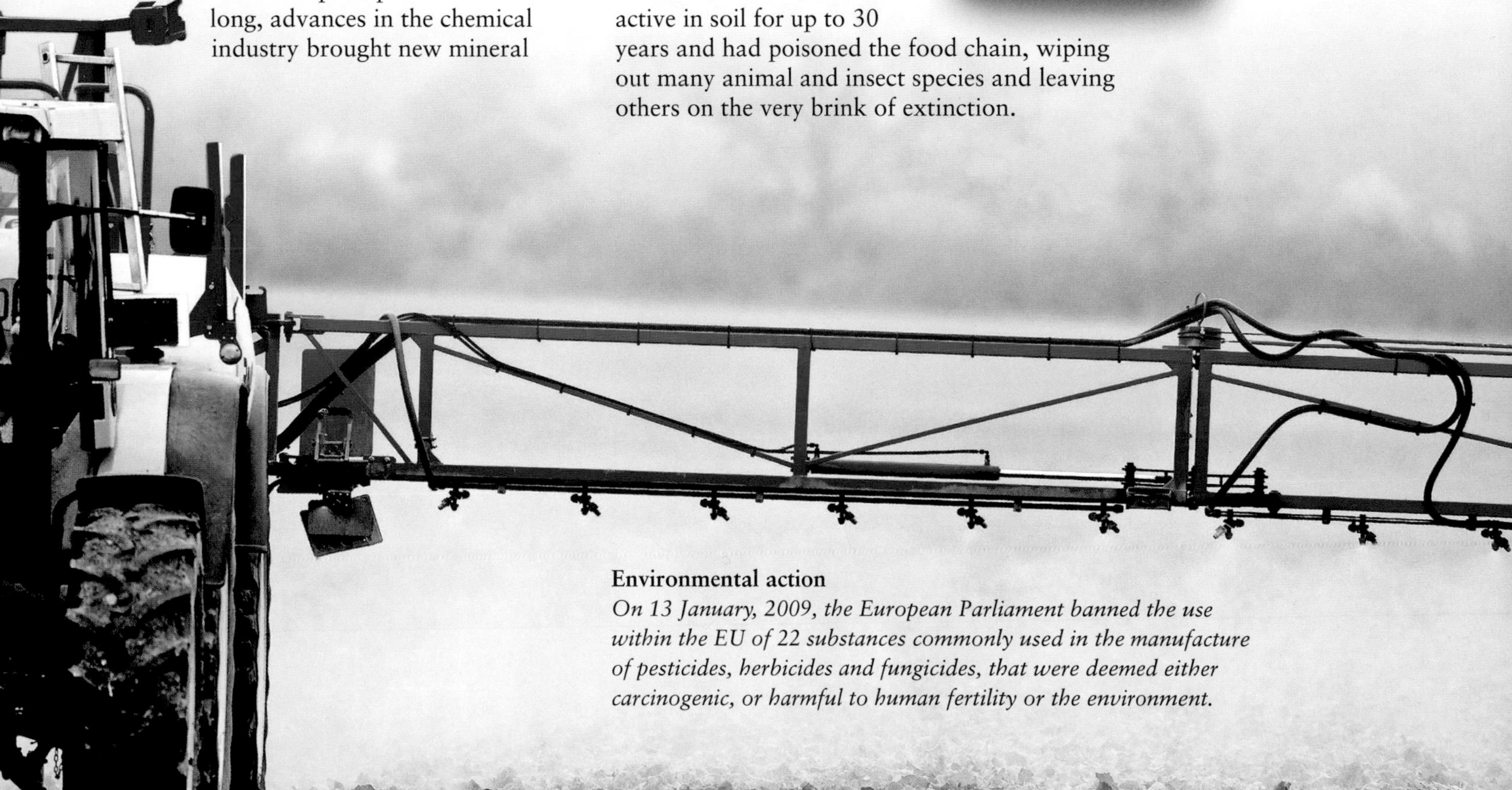

Environmental action
On 13 January, 2009, the European Parliament banned the use within the EU of 22 substances commonly used in the manufacture of pesticides, herbicides and fungicides, that were deemed either carcinogenic, or harmful to human fertility or the environment.

JOSEPH PRIESTLEY – 1733 TO 1804

Radical man of God and science

Joseph Priestley was a brilliant chemist who specialised in the study of gases. He was responsible for a whole series of important discoveries, from oxygen to plant transpiration. Yet his radical religious stance – he was a nonconformist minister – and his support for the French Revolution severely compromised his scientific career and eventually forced him to leave England for good.

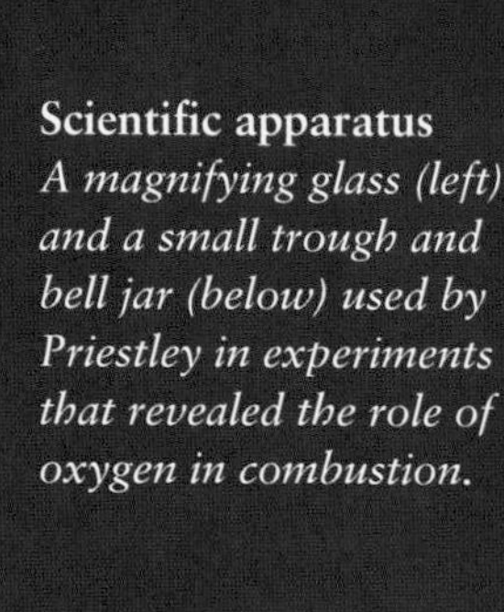

Scientific apparatus
A magnifying glass (left) and a small trough and bell jar (below) used by Priestley in experiments that revealed the role of oxygen in combustion.

Local hero
A statue of Joseph Priestley (below), by the sculptor Alfred Drury, was erected on City Square in Leeds in 1899.

One rainy day in August 1774, Joseph Priestley, then 41 years old, was immersed as usual in his chemistry experiments. Although he did not realise it, he was about to make history. The experiment he was engaged in was an attempt to corroborate the phlogiston theory, the widespread but erroneous hypothesis – scientific orthodoxy in the 18th century – that all combustible substances contained a fluid known as 'phlogiston', which was set free when the substance was burned.

Priestley placed a piece of red mercuric oxide under a glass bell jar, focused a ray of sunlight on it through a magnifying glass and set it alight. As it burned, the mercuric oxide changed into shiny globules of metallic mercury and released a colourless gas. To find out what the gas consisted of, Priestley placed a lit candle under the bell jar, expecting the phlogiston to 'saturate' the air under the glass and snuff out the flame. Instead, the candle burned 'with a remarkably vigorous flame'. Here, Priestley the phlogistonist drew the wrong conclusion, reasoning that the gas which allowed the candle to burn brightly must be 'dephlogisticated air'. Though he failed to realise it, he had just exploded the phlogiston theory and discovered the most abundant gas on Earth: oxygen.

Priestley's experiment was repeated a few months later by the French chemist Antione Laurent de Lavoisier, an opponent of the phlogiston theory. Lavoisier suspected that this mysterious new gas, which he dubbed 'oxygen', played a key role in combustion.

A gifted student

There was nothing in Priestley's background to suggest that he was destined for scientific greatness. He was born in 1733 in Birstall Fieldhead near Leeds. His father was a strict Calvinist, who made a modest living as a textile worker. When his mother died prematurely, his father remarried and Joseph went to live with a wealthy and childless uncle and aunt, who practised a more liberal form of Calvinism. The precocious young Joseph thrived in this educated, free-thinking atmosphere and soon developed a lively, independent mind. Preparing himself for the ministry, he mastered 'Hebrew, Chaldee, Syriac, and a little Arabic'.

At the age of 22, Priestley was appointed assistant minister to a Presbyterian congregation in Needham Market, Suffolk,

DEDICATED TEACHER

Priestley was passionate about education, introducing into his own curriculum new practical courses on the sciences and modern history. Throughout his life, he published papers on a wide range of subjects, including grammar, educational theory, politics, history and religion. He took a keen interest in rationalist philosophy, particularly the question of human free will, and was firmly convinced that 'children can be moulded to be whatever we wish them to be'. Priestley was also far ahead of his time in arguing that women should be allowed to continue their education beyond school level.

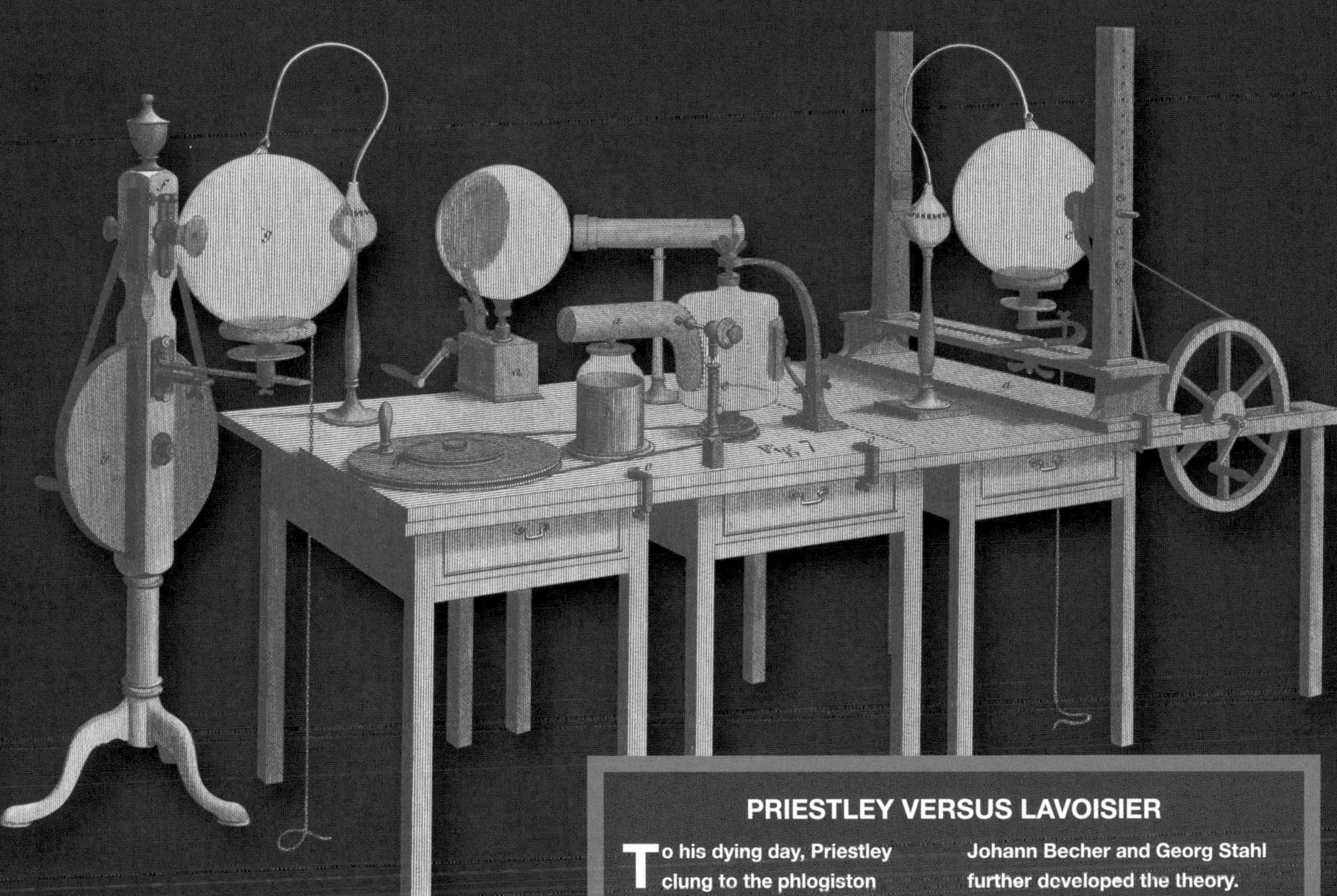

PRIESTLEY VERSUS LAVOISIER

To his dying day, Priestley clung to the phlogiston theory, first set out by Aristotle: 'Phlogiston is a fire-like substance contained within all matter, which is released during the process of combustion'. The Arab alchemist Jabir ibn Hayyan (8th–9th centuries) and the 17th-century German chemists Johann Becher and Georg Stahl further developed the theory. Priestley's views put him at odds with Lavoisier, who believed that combustion resulted not from the liberation of 'phlogiston' but from the reaction of 'eminently respirable air' (oxygen) with the metal or organic matter of the substance being burned.

and six years later, in 1761, he became a tutor at Warrington Academy in Lancashire, teaching literature, anatomy, astronomy and botany. He discovered he had a real gift for devising scientific experiments, particularly those involving electricity. As his reputation grew he was elected to the Royal Society in 1766 and he struck up a friendship with Benjamin Franklin, who encouraged him to publish his first scientific treatise, *The History and Present State of Electricity*, in 1767. In that treatise he correctly posited that charcoal conducts electricity. The same year, he resigned his post in Warrington and became minister of Mill Hill Chapel in Leeds. Then, in 1772, William Fitzmaurice-Petty, 2nd Earl of Shelburne, appointed Priestley as librarian and tutor to his sons. Shelburne had a private laboratory on his estate in Wiltshire, which was to become the scene of some remarkable discoveries over the next six years.

Master experimenter

Priestley's investigations of the properties of gases made him a pioneer of pneumatic chemistry. Using a 'pneumatic trough' filled with mercury, he collected and isolated no fewer than ten water-soluble gases, including carbon monoxide, sulphur dioxide, ammonia and nitrous oxide. In 1774 he published his findings in *Experiments and Observations on Different Kinds of Air*. By now, Priestley was known throughout Europe.

Along the way, quite by chance, Priestley discovered a practical method of carbonating water. He suspended a carboy of water over a vat of fermenting grain at a brewery in Leeds and found that the rising carbon dioxide infused the water. The resulting 'soda water' had a pleasant, slightly acidic taste like that of the natural mineral waters then much in vogue for their curative properties.

In 1772 Priestley made another crucial discovery, which paved the way for a fuller understanding of photosynthesis. He found that a mouse placed under a bell jar containing some sprigs of mint did not suffocate, whereas

Unwieldy but effective
Priestley used this curious-looking contraption (above) to demonstrate that charcoal conducts electricity. The resulting work, The History and Present State of Electricity, *laid down some fundamental principles of electrical forces, including the fact that they follow an inverse square law.*

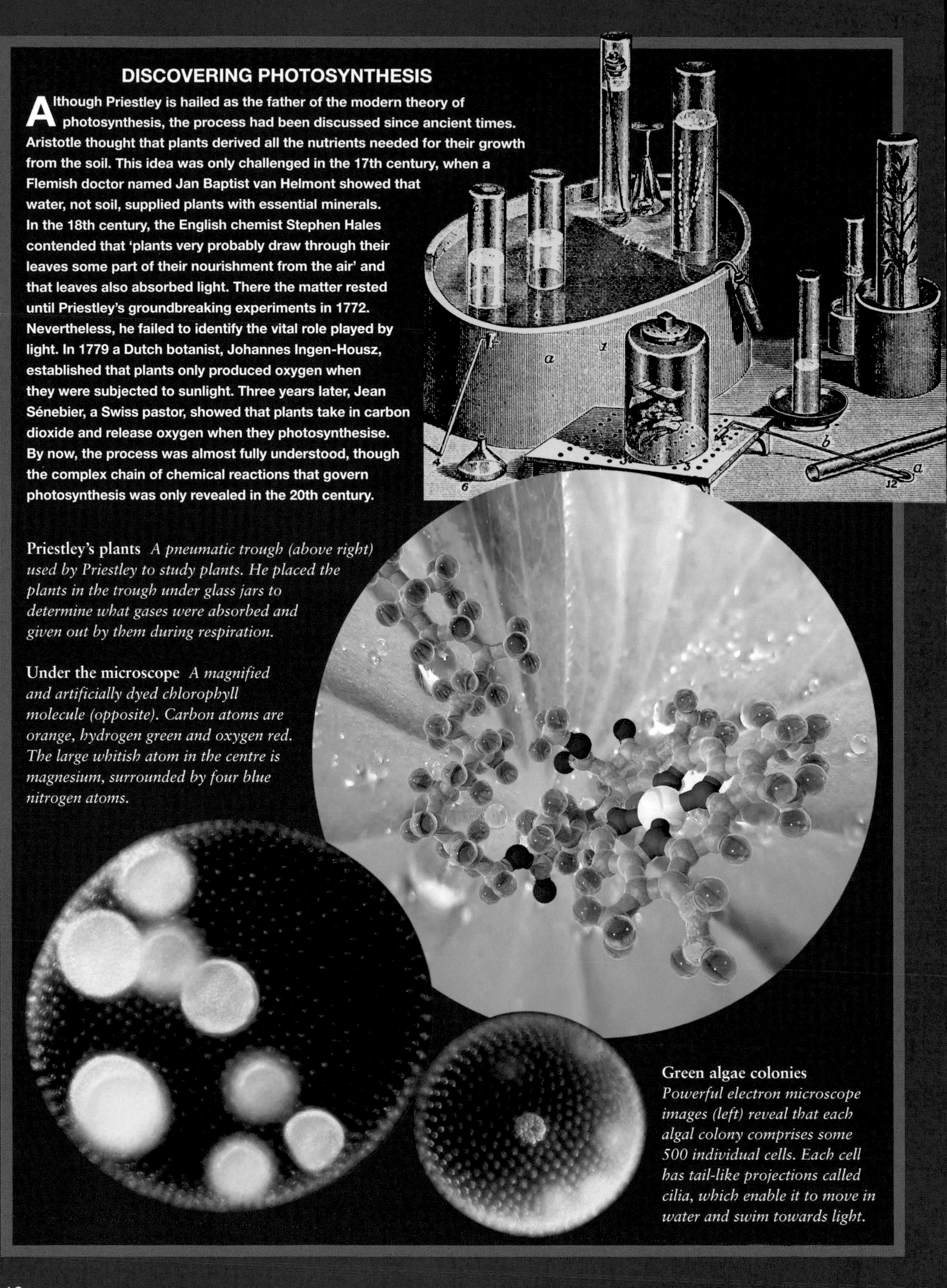

DISCOVERING PHOTOSYNTHESIS

Although Priestley is hailed as the father of the modern theory of photosynthesis, the process had been discussed since ancient times. Aristotle thought that plants derived all the nutrients needed for their growth from the soil. This idea was only challenged in the 17th century, when a Flemish doctor named Jan Baptist van Helmont showed that water, not soil, supplied plants with essential minerals. In the 18th century, the English chemist Stephen Hales contended that 'plants very probably draw through their leaves some part of their nourishment from the air' and that leaves also absorbed light. There the matter rested until Priestley's groundbreaking experiments in 1772. Nevertheless, he failed to identify the vital role played by light. In 1779 a Dutch botanist, Johannes Ingen-Housz, established that plants only produced oxygen when they were subjected to sunlight. Three years later, Jean Sénebier, a Swiss pastor, showed that plants take in carbon dioxide and release oxygen when they photosynthesise. By now, the process was almost fully understood, though the complex chain of chemical reactions that govern photosynthesis was only revealed in the 20th century.

Priestley's plants *A pneumatic trough (above right) used by Priestley to study plants. He placed the plants in the trough under glass jars to determine what gases were absorbed and given out by them during respiration.*

Under the microscope *A magnified and artificially dyed chlorophyll molecule (opposite). Carbon atoms are orange, hydrogen green and oxygen red. The large whitish atom in the centre is magnesium, surrounded by four blue nitrogen atoms.*

Green algae colonies *Powerful electron microscope images (left) reveal that each algal colony comprises some 500 individual cells. Each cell has tail-like projections called cilia, which enable it to move in water and swim towards light.*

it would if the vegetation was removed. He noted: 'I have discovered a method of restoring air that has been vitiated [impaired] ... and also found at least one of the principal means of restoration that nature employs to this same end, namely vegetation'.

In addition to discovering oxygen, Priestley isolated ammonia ('alkaline air') and studied its decomposition when an electrical current was passed through it. And together with Henry Cavendish, he established that water was a compound (hydrogen + oxygen) and successfully synthesised it as H_2O.

Driven into exile

Alongside his scientific works, Priestley also published numerous theological tracts, in which he questioned many of Christianity's most cherished beliefs, including the Trinity, predestination and the divine origin of the Bible. His views made him powerful enemies among the political and religious establishment. Then, in 1791, Priestley published *Letters to the Right Honourable Edmund Burke*, in which he voiced his strong support for the French Revolution. In response a mob attacked his home, Fair Hall in Birmingham, and razed it to the ground. His laboratory was destroyed and many of his papers burned or lost.

Priestley fled in disguise to Worcester, then to London, then France. He became a French national in 1792 and emigrated to the United States in 1794, joining his two sons who had settled in Philadelphia. He led a frugal life, devoting his time to study, and died in 1804. Thomas Cooper, a close companion in his final years, wrote of Priestley: 'He kept on writing right to the end of his life, and was cheerful and positive to the last.'

PRIESTLEY THE 'LUNATICK'

During his time in Birmingham, Priestley was a member of the Lunar Society, a group of scholars, scientists, engineers and free-thinkers that was effectively a 'think tank' for the Industrial Revolution. Together with other Society luminaries – John Wilkinson, Matthew Boulton, James Watt, Erasmus Darwin (grandfather of Charles), James Keir – Priestley campaigned for the practical application of scientific innovation in industry, transport, education and medicine.

Figure of hate
Priestley was a favourite subject for caricaturists in Britain, who not only lampooned him as 'Dr Phlogiston' but also portrayed him as a crazed political agitator trampling the Bible underfoot (above). Incited by such scurrilous attacks, a mob burned down his home in 1791 (top).

BRAKES – *c*1767

Reining in speed

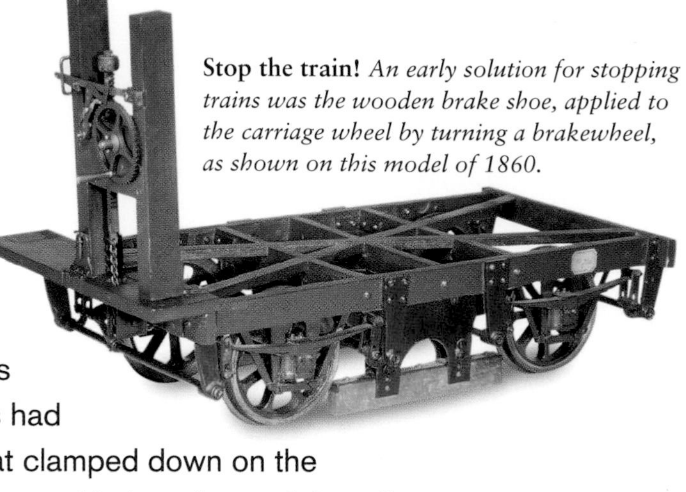

Stop the train! *An early solution for stopping trains was the wooden brake shoe, applied to the carriage wheel by turning a brakewheel, as shown on this model of 1860.*

When carts and carriages were pulled by draught animals, stopping was simple. But as transport technology grew ever more complex, mechanical means of stopping were needed and brakes were invented. Even in the 17th-century, some coaches had crude brakes – wooden blocks, operated by a lever, that clamped down on the wheel rims. But the story of brakes really begins in earnest with the advent of the railway.

The earliest ancestor of the railway is thought to have been a tramway of wooden rails built at a silver mine at Leberthal, in Alsace, in around 1550. Small wagons were loaded up with ore and pushed along the rails. In several sections, a sloping gradient allowed the wagons to trundle along sedately under their own momentum. Braking only became an issue after 1767, when foundryman Richard Reynolds first introduced iron rails into the coalmines at Coalbrookdale, cradle of Britain's nascent Industrial Revolution in Shropshire. These new rails had a U-shaped cross-section for the carts' wheels and allowed wagons to run much faster than wooden rails. In consequence they were fitted with metal brake shoes worked by levers to stop them.

A dangerous occupation

The only means of slowing the very first locomotives, in the early 1800s, was to reverse the steam flow to the cylinders. This was not only inefficient, it also damaged the wheel treads, and before long it was supplanted by brake shoes on the wheels. Train drivers were responsible for braking, a job that required an extremely strong arm – even after the original lever system had been replaced by a hand-operated brakewheel and screw arrangement.

As locomotive speeds increased, it became necessary to fit brakes to each carriage or wagon of the train and the job of 'brakeman'

THE WORLD'S FIRST AUTOMOBILE ACCIDENT

The first automobile accident in history took place on 23 November, 1770, during a demonstration of a 'steam-carriage' (*fardier à vapeurs*) by its inventor Nicolas-Joseph Cugnot at the Arsenal in Paris. The vehicle had no brakes and could not be prevented from demolishing a wall (above). More than a century later, the motor car took to the road, once more raising the question of how to bring it to a safe stop. There was no possibility of using brake blocks on wheels rimmed with either solid rubber or pneumatic tyres. Instead, engineers came up with the idea of a braking system not on the wheel itself but on a metal component – a disc or drum – mounted on the axle. To slow and stop the wheel, brake callipers, originally operated by either rod linkages or chains, pressed brake pads against the sides of the disc or drum. This same principle was retained even after mechanical brake-control systems were overtaken by hydraulic brakes from 1918 onwards (patented by Malcolm Loughead). Servo-assisted brakes, in which a pump enables drivers to apply greater braking force, appeared shortly after the end of the Second World War.

STOPPING THE SAILS

Watermill wheels could easily be slowed by diverting the flow of water upstream of the millrace, but windmills were a different matter. The first brakes ever made were for windmills. They comprised a flexible leather sleeve, worked by a lever, which forced wooden blocks down onto the gear wheel at the head of the main drive shaft to stop it turning. Variations on this basic system were used up to end of the 14th century, when the 'Flemish Brake' – a compressible metal ring almost entirely surrounding the gear wheel – began to supplant earlier mechanisms.

emerged. On a signal from the driver (usually a blast of the whistle), the brakeman had to leap from carriage to carriage and screw all the brakewheels shut. It was dangerous work and accidents caused by misunderstandings between brakemen and drivers were common.

From the mid-19th century onwards, much thought was given to devising methods of continuous braking – that is, a system controlled solely by the driver, which slowed the whole train simultaneously. A number of different systems were patented, using a variety of technologies, including electromagnets, steam, vacuums and compressed air.

The Westinghouse brake

The system that won out harnessed compressed air in a device unveiled in 1869 by the American engineer and entrepreneur George Westinghouse. The train driver opened a valve on the locomotive which pumped compressed air into a hose running the entire length of the train. This hose, or brake pipe, was connected to a cylinder on each carriage containing a piston connected to the brake shoes. When the pipe was filled with compressed air, the piston activated the brakes shoes, or put the brakes on – a 'direct' form of air brake. But the system had two major drawbacks. Firstly, it took time to repressurise the system after each braking. Secondly – and most importantly – the brakes failed if any part of the hose was split.

Three years later, Westinghouse introduced the 'triple-valve air brake' to solve these flaws. This worked on the opposite principle to his original system: the default setting for the brakes – that is, when the hose was unpressurised – was 'on' and compressed air was forced into the pipe to release them. A decrease in pressure, either as a result of the driver releasing the valve, or due to a leak or accidental decoupling, would cause the brakes to re-engage. This highly responsive, fail-safe system was adopted by railways worldwide, making a fortune for the inventor and his firm, the Westinghouse Air Brake Company (WABCO).

Spectacular brake failure
On 22 October, 1895, the Granville to Paris express overran the buffer stop and smashed through the façade of Montparnasse station in Paris. The train's Westinghouse brake was found to be faulty, but the excessive speed of 63km/h (39mph) as the train entered the station was also blamed for the crash.

Roll up, roll up!
The circus founded in 1907 by Carl Hagenbeck and Benjamin Wallace toured the USA, displaying animals from the four corners of the Earth. This colourful promotional poster is from 1912.

Circuses 1768

In 1768 a former sergeant-major in the King's Light Dragoons named Philip Astley set up a riding school on a piece of open ground south of Westminster Bridge in London. Equestrian shows were popular at the time and he began to put on displays of horsemanship for the general public.

Astley's audience crowded around a circular, roped-off arena to watch him perform acrobatic feats on horseback. At one point in the routine, his horse lay down to play dead. Though the shows originally took place in the open-air, before long Astley had a permanent amphitheatre with a roof and was attracting spectators from every social class. George III even asked him to stage a private performance for the court. The term 'circus' was coined by a rival of Astley's, Charles Hughes, who opened the 'Royal Circus and Equestrian Philharmonic Academy' in 1782.

Animals, trapeze artists and clowns

The new form of entertainment soon spread to France, Russia and the United States. The 1820s saw the appearance of the 'Big Top', a large tent that could be put up and taken down to allow the circus to move from town to town. Meanwhile, circus programmes, boldly advertised on colourful posters, kept changing to provide a stream of novelties. One of the first circus elephants, Old Bet, was exhibited in 1815 by the American showman Hachaliah Bailey. A Frenchman, Jules Léotard, invented the flying trapeze in 1859, while in 1888 the German animal trainers Wilhelm and Carl Hagenbeck introduced the circus cage for displaying wild animals in the ring. Around the same time, the double-act of a whitefaced clown and a mischievous red-nosed clown – parodying authority and rebellion – began to appear. The classic comedy pairing of straight man and funny man was born. In 1880, the American entrepreneur Phineas Taylor Barnum introduced three rings, so three different acts could perform at the same time.

In its heyday the circus produced famous dynasties such as the Chipperfields who toured the world, but after the 1970s the circus went into steep decline. In recent years it has enjoyed a revival with new troupes like the Canadian 'Cirque du Soleil', who breathed new life into the format with an exciting new approach to lighting, music, costume and above all the acts themselves.

Circus pioneer
Philip Astley's 'Amphitheatre' in a contemporary print by the artist Thomas Rowlandson (1808).

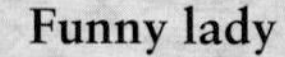

Funny lady
Annie Fratellini (1932–97) was the first woman in France to take on the role of the 'auguste' clown with the red nose. She went on to found the French national circus school in Paris in 1972.

GREATEST SHOWMAN ON EARTH

The legendary American impresario Phineas T Barnum (1810–91) entered the circus business late in life, in 1872. Never a man to do things by halves, he billed his extravaganza – which featured a travelling circus, a menagerie and a sideshow of freaks – as 'The Greatest Show on Earth'. The show's star attraction was a huge African elephant, Jumbo, purchased by Barnum in 1882, but sadly killed in 1885 charging a train in Canada.

MODERN CHEMISTRY – 1777

From alchemy to science

Chemistry – the science that studies the combinations of atoms and molecules in matter and the reactions between them – became a modern discipline after the French scholar Antoine-Laurent de Lavoisier managed to distil the complex rules governing chemical reactions into a few basic guiding principles.

'As dangerous as is the desire to systematise in the physical sciences, it is, nevertheless, to be feared that in amassing, in no particular order, a great multiplicity of experiments we obscure the science rather than clarify it, render it difficult of access to those desirous of entering upon it, and finally, obtain at the price of long and tiresome work only disorder and confusion ... Encouraged by these reflections, I venture to propose to the Academy today a new theory of combustion, or rather, to speak with the reserve which I customarily impose upon myself, a hypothesis by the aid of which we may explain in a very satisfactory manner all the phenomena of combustion and of calcination, and in part even the phenomena which accompany the respiration of animals.'

Thus Antoine Lavoisier introduced his latest treatise to the French Royal Academy of Sciences – the paper is dated 5 September, 1777, but was presented to the Academy on 12 November that year. It was entitled *Sur la combustion en général (*'On combustion in general)' and it signalled nothing less than the birth of modern chemistry.

Building blocks of a new discipline

In barely nine pages of text, Lavoisier set out the basis of a new scientific discipline. Chemistry had been known by the name since the mid-17th century, but until now it had been little more than a hotchpotch of disparate and often contradictory theories on the way in which various substances fused and combined to form the diverse physical matter we encounter on Earth. It took him almost two more decades – until his untimely death by guillotine in 1794, during France's reign of terror – to corroborate his theory, all the salient points are evident in this brief paper.

On the one hand, Lavoisier contended, all matter is ultimately an assemblage of simple chemical elements (atoms). On the other, new matter can only be produced through the combination of already existing matter – in other words, nothing is created and nothing

Jottings of a genius
The notebook used by Lavoisier for his research into the decomposition of water in 1768.

LAVOISIER'S FAMOUS DICTUM

When formulating his Law of the Conservation of Mass in chemical reactions, Lavoisier is commonly believed to have coined the maxim 'Nothing is created, nothing is lost, everything is transformed'. In fact, this is a condensed version of a passage written in his *Elementary Treatise on Chemistry* (1798): 'Nothing is created, neither in the process of art nor of nature, and in principle it can be stated that in every process there is an equal quantity of matter before and after it has taken place, that the quantity and quality of the principles are the same, and that there are only changes, modifications'.

INVESTIGATING COMBUSTION

Lavoisier expounded on his radical new approach to chemistry in his masterpiece, the *Elementary Treatise on Chemistry*, which was published four years after his death. The work was the fruit of intensive research into the nature of combustion, which he had conducted between 1773 and 1777. Taking what he could from others in the field, notably Joseph Priestley and Henry Cavendish, Lavoisier's main apparatus for his experiments consisted of a glass bell jar resting in a dish of water, which sealed off the air inside from the external atmosphere. He used this equipment to see what happened to a lit candle placed within the closed environment. His conclusion was that ordinary air must contain an element – which he called 'oxygen' – responsible for combustion, and that the heat of the flame causes it to combine with another element (carbon) to produce a third element (carbon dioxide, or CO_2) that is non-combustible. He also noted that the total volume of air was reduced by the reaction, and that the water level rose under the bell once the flame went out. This did not, he claimed, indicate that the total quantity of matter had changed as a result of the reaction, but simply that the new gas took up less space than air.

Combustion chamber
The world's first ice-calorimeter, used in 1782–3 by Lavoisier and Pierre-Simon Laplace to examine the heat generated by various chemical reactions.

Investigating decomposition
A piece of apparatus devised by Lavoisier for use in experiments to investigate putrid fermentation (below). This phenomenon, too, bore out his Law on the Conservation of Matter.

disappears in chemical reactions. Moreover, the reactions that produce matter either require heat in order to take place or give off heat.

The concepts outlined in Lavoisier's paper may seem fundamental today, but at the time they were not remotely self-evident. They included the idea that atoms form the basic building blocks of matter; the fact that the incredible diversity of matter on Earth derives from a finite (and extremely limited) number of types of atom; the notion that nothing is created or lost when a chemical reaction takes place; and the insight that heat is not a chemical substance. As was the case with all of the sciences, it had taken several centuries and many discoveries, plus a gradual, barely perceptible paradigm shift in mindset, for one thinker to emerge who offered a clear explanation of nature's innermost workings.

The atoms of Democritus

Modern chemistry is based primarily on the notion of simple elements known as atoms. Yet the atomic, or atomistic, theory of matter is actually more than 2,000 years old. Two ancient Greek philosophers first came up with the idea: Leucippus and his pupil Democritus (*c*460–*c*370 BC). Their hypothesis, partly scientific and partly metaphysical, was a refutation of the ideas put forward by Aristotle and his followers, who regarded matter as a continuum composed of the four classical elements (fire, water, earth and air), existing in various proportions and infinitely divisible. By contrast, Leucippus and Democritus

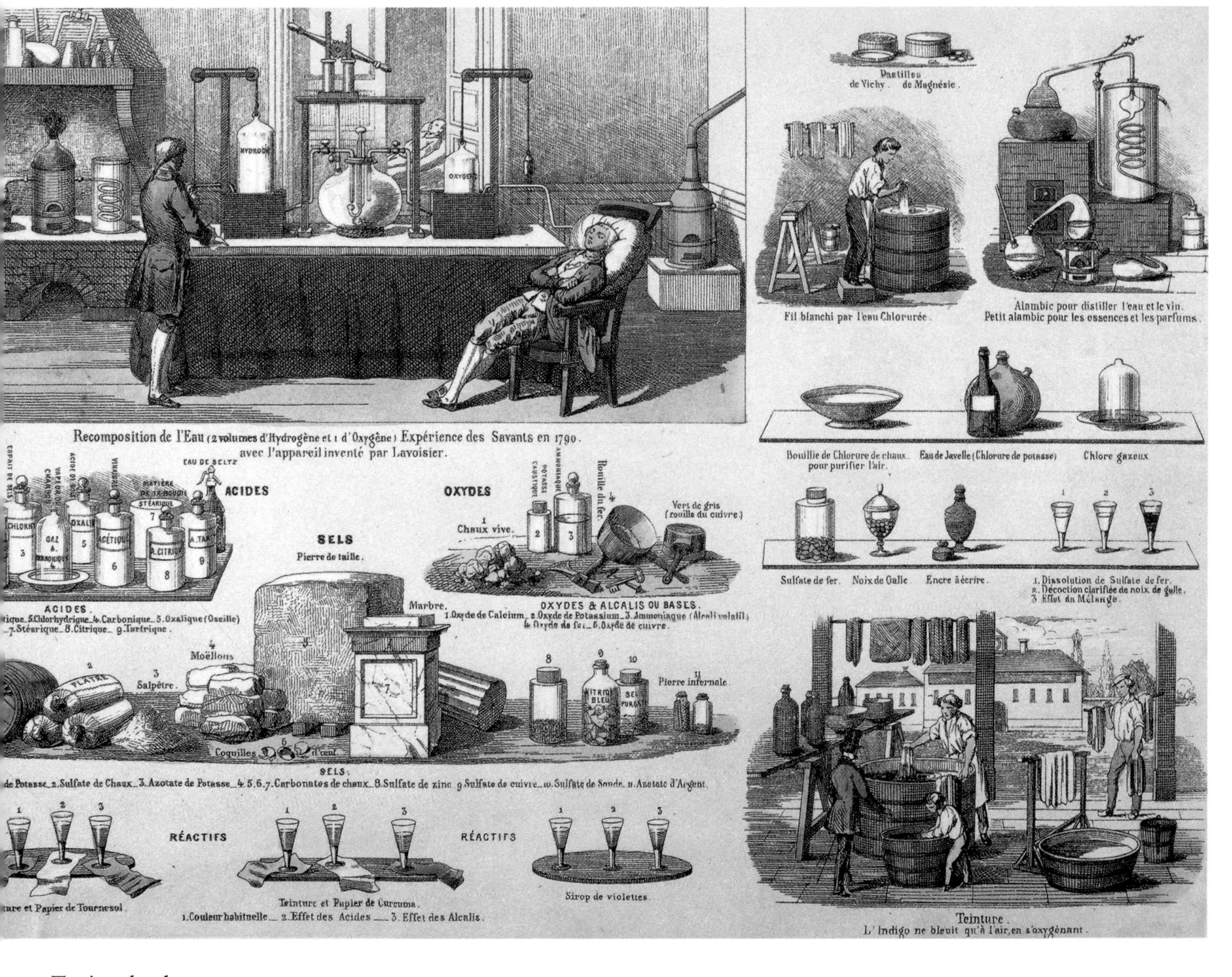

Testing the theory

A 19th-century engraving shows the various stages of experimentation that Lavoisier conducted to reconstitute water from hydrogen and oxygen.

HELIUM – THE GAS OF THE SUN

Among all the new discoveries made by chemists in the 19th century, a special place was reserved for the gas helium. For this element was found not on Earth but on the Sun. By this stage, chemists knew that, when heated, every element emitted a particular colour, known as a 'spectral ray'. This insight came as a result of the invention of the spectrometer, an optical instrument that split light into its constituent wavelengths. In 1868, the French astronomer Pierre Janssen was analysing the Sun's light spectrum and spotted a ray that resembled no known element. He mentioned it to his British colleague Norman Lockyer, who concluded that it was produced by a new element, which he named 'helium' after the Greek word for the Sun.

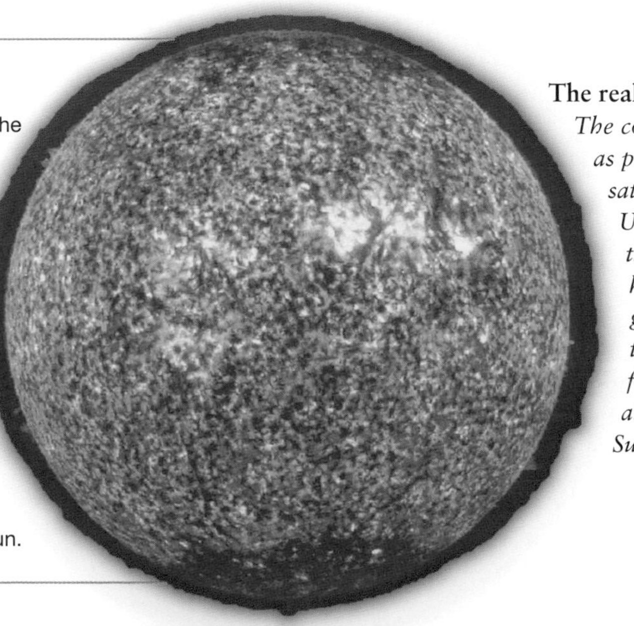

The realm of Helios

The colour of the Sun, as photographed by a satellite through a UV filter, reveals the presence of helium. This inert gas is produced by thermonuclear fusion of hydrogen atoms in the Sun's core.

maintained that matter was 'discontinuous', being made up of tiny individual and indivisible particles – atoms – that join together to form more substantial bodies.

Battling Aristotle

Yet the Aristotelian concept of the four elements won out and continued to hold sway over scientific thought, not only in ancient Greece but also in the Arab world and in Europe, from the 4th century BC to the 17th century AD. Then, in 1620, the English polymath Francis Bacon – a great aficionado of ancient Greece and an anti-Aristotelian – revived the atomic theory of Leucippus and Democritus, which had also been expanded by Epicurus. Bacon was closely followed by the French philosopher and mathematician Pierre Gassendi, who in 1624 redefined the theory, stripping it of its metaphysical elements.

This new atomic theory became widely accepted; in a work of 1674 entitled *Cours de Chymie* ('Chemistry Course'), for example, the French pharmacist Nicolas Lémery put forward the idea that certain substances can be broken down into their smallest constituent parts, namely atoms. Over the course of the 18th century a succession of thinkers attempted to describe atoms, detailing the various types, their forms and properties, to determine the laws that governed their combination, and to isolate and study these basic units of matter, which were still poorly understood.

Fine craftsmanship
A set of chemical scales belonging to Lavoisier (right). He had his instruments specially made by the most skilled craftsmen of the age, to ensure the greatest possible accuracy of measurement.

Disproving the phlogiston theory

Towards the end of the 17th century, the prevailing orthodoxy on combustion was the phlogiston theory. This contentious hypothesis sparked off a fierce debate concerning the chemical composition of matter. The basic proposition of the phlogiston theory, as put forward by its chief exponents, the German chemists Georg Ernst Stahl and Joachim Becher, was that a fire-like substance called phlogiston was present in all flammable matter, regardless of the individual make-up of the substance, and that this was released during combustion.

In the decade 1760–70, chemistry pioneers like Henry Cavendish and Joseph Priestley attempted to isolate different parts of air, such as 'phlogisticated air', 'dephlogisticated air' and 'inflammable air' – and believed that they had succeeded. Surrounded by the phlogiston forest, they failed to see the new wood of truth emerging from their experiments. The notion of phlogiston – a fluid that supposedly contained the latent heat within all bodies – remained unassailable right up to 1777, when Lavoisier neatly refuted the theory in his brief paper on combustion: 'Now if we demand of the partisans of the doctrine of Stahl [the phlogiston theory] that they prove the existence of the matter of fire in combustible bodies, they necessarily fall into a vicious circle and are obliged to reply that combustible bodies contain the matter of fire because they burn, and that they burn because they contain the matter of fire. It is easy to see that in the

NAMING THE ELEMENTS

Oxygen, hydrogen, carbon, nitrogen ... by the end of the 18th century, chemists had succeeded in identifying around 50 different types of atom. By the 19th century, the list of chemical elements had grown to 92. In 1803 the English chemist John Dalton published a table showing the relative atomic weights of elements, using for the first time a system of symbolic notation: O = oxygen, C = carbon, N = nitrogen. This laid the groundwork for a systematic classification of the elements. Several types of classification were proposed, but it was the Russian chemist Dmitri Ivanovich Mendeleev who, in 1869, came up with the periodic table of chemical elements (below). In drawing up the table Mendeleev was so accurate and farsighted that he even anticipated the existence of elements as yet undiscovered and left blank spaces for them.

Dmitri Ivanovich Mendeleev (1834–1907)
Elements discovered during Mendeleev's lifetime included gallium (Ga) and scandium (Sc)

RISE OF THE ATOM

After Lavoisier, new discoveries in the world of chemistry came thick and fast. His conclusions had shown that the chemical transformation of matter involved the union or separation of different types of atom, each of which was known as a 'chemical element', such as hydrogen, oxygen, carbon, etc. Over the following century, chemists found that each of these elements had a specific atomic weight. Henceforth, their research was aimed at isolating new elements and determining their weight, or atomic mass, and trying to establish a link between the mass of an element and the way it combined with others to form molecules. Eventually, in attempting to construct a model of the atom, scientists discovered at the turn of the 20th century that it was far from indivisible. This marked the beginning of the theory of elementary particles and quantum physics.

last analysis this is explaining combustion by combustion.' Lavoisier went on to banish phlogiston from chemistry by demonstrating that combustion of a flammable material does not result in the release of one particular element that is always identical – that is, phlogiston – but rather that combustion breaks down the substance being burned into its constituent elements.

This sounded the death knell for the phlogiston theory. The chemical reactions in combustion can be explained as an interplay of union and separation between basic units of matter, such as oxygen, hydrogen and carbon, which is triggered by heat being exchanged between them. Lavoisier's radical theory of combustion truly paved the way for modern chemistry.

Paradigm shift
In the early 20th century, the discovery of the composition of the atom, which is made up of electrons spinning around a central nucleus, overturned the foundations of chemistry once more, opening up new fields of study such as particle and quantum physics.

ANTOINE-LAURENT DE LAVOISIER – 1743 TO 1794

Master craftsman of the chemical revolution

A highly cultured man, gifted with great mental acuity and a meticulous eye for detail in devising experiments, Lavoisier is acclaimed as one of the founding fathers of modern chemistry. By a cruel irony of fate, the same Age of Enlightenment that made his groundbreaking work possible also spawned the French Revolution which claimed his life. Lavoisier was tried and condemned during the Reign of Terror and guillotined in Paris in 1794.

Enemy of the state *Lavoisier's role as a tax collector brought the wrath of revolutionary zealots down on his head, and he was arrested in his laboratory in 1793.*

'The Republic has no need of scientists or chemists; justice must be allowed to run its course.' With this brusque and extraordinarily short-sighted sentence, the president of the Revolutionary Tribunal, Jean-Baptiste Coffinhal, brushed aside Lavoisier's plea for clemency. The date was 8 May, 1794, and the eminent 50-year-old chemist found himself convicted of high treason, together with 27 of his fellow *fermiers généraux* (private tax collectors). Later that same day, he was guillotined on the Place de la Révolution (now the Place de la Concorde) in Paris. Like so many of his fellow citizens during the Reign of Terror (September 1793 to July 1794), Lavoisier found himself falsely accused, summarily tried and swiftly executed. As his friend Louis Lagrange lamented: 'It took them only an instant to cut off his head, but France may not produce another like it in a century.'

A thirst for knowledge

Antoine-Laurent de Lavoisier was born on 26 August, 1743, into an aristocratic family in the suburbs of Paris. His mother died in 1748 and the young Antoine inherited her fortune, which later gave him free rein to pursue his studies. It soon became clear that the young man had an aptitude for learning.

AN UNPOPULAR ROLE

After qualifying as a lawyer in 1764, Lavoisier followed a well-trodden path for men of his class, investing part of his fortune in the *Compagnie des fermiers généraux*, an excise and tax-collecting institution backed by large amounts of private capital (equivalent to around 3 billion euros today). By purchasing shares, investors in the company gained the right to levy taxes on a wide range of goods and services in France, such as customs duty, property tax and even a salt tax. They paid a portion of this over to the Royal Treasury, but kept the rest for themselves. Not surprisingly, the huge remuneration they extracted for themselves from the populace did not make the *fermiers* popular with the public at large. For Lavoisier, as well as producing income, the role helped him to hone his administrative and political skills and also stimulated his interest in economics, on which he wrote several papers, and he remained active in the company right up to his arrest.

At his father's instigation he began to study for a law degree at the University of Paris in 1761. At the same time he enrolled on courses in physics, botany, medicine, mineralogy, geology and chemistry, which were being taught by leading lights in those fields.

A wide-ranging intellect

In 1768, at the age of 25, Lavoisier was elected a member of the French Royal Academy of Sciences. He was commissioned to submit a proposal for street lighting in Paris and to draw up a mineralogical map of France. The young Lavoisier was fascinated by all aspects of science, working in such disparate areas as the biology of snails, the problem of noxious gases rising from septic tanks, geology and optics. Even so, he devoted most of his intellectual energy to investigating the nature of air. From 1772–3 onwards, he began to take a close interest in the properties of air, investigating its role in animal respiration as well as in combustion. He spent 20 years making a detailed study of the work of his predecessors in the field. He accepted some of their findings but rejected others, and published extensively on the subject. Aided by his wife, Marie-Anne Pierrette, he transformed his house into a state-of-the-art laboratory, where he conducted a series of famous experiments that ultimately defined the foundations of chemistry.

Captured in bronze
A statue of Lavoisier by French sculptor Aimé-Jules Dalou (1892).

Devoted to science
A portrait of Lavoisier and his wife Marie-Anne Pierrette, painted in 1788 by the French Neoclassical artist Jacques-Louis David.

UNSUNG HEROINE

When Lavoisier joined the *Ferme générale* in 1771, he made such a good impression on his boss, Jacques Paulze, that Paulze offered the young man the hand of his 13-year-old daughter, Marie-Anne-Pierrette. In time, she became Lavoisier's closest collaborator in his scientific endeavours. She helped him set up his laboratory, invented new pieces of apparatus and translated Joseph Priestley's work from English. Despite the prevailing prejudice against educated women, she became a renowned chemist in her own right. After her husband's death, Madame Lavoisier made it her life's work to edit, publish and translate his writings.

Iron and steel bridges 1779

Elegant crossing
The Iron Bridge (below) has spanned the River Severn since 1779. Reinforced concrete has now largely replaced iron and steel in many of the world's iconic new bridges.

Spanning the River Severn as it flows through Shropshire is an historic piece of engineering. The Iron Bridge – 60m long and 30m high (196ft and 98ft) – has stood on its site at Coalbrookdale for 230 years, the first metal bridge ever built in the world. The bridge was designed by the architect Thomas Pritchard and manufactured by the ironmaster Abraham Darby III, grandson of the man of the same name who invented the process of smelting in a blast furnace fuelled by coke.

In taking on this daunting venture, Darby was opening up a new market for the cast iron produced by his firm, the Coalbrookdale Company. Erection of the bridge began in 1776; its basic skeleton comprised five iron ribs forming a semicircular arch. Some 800 massive iron components were individually cast and then assembled using traditional mortise-and-tenon and dovetail joints, following tried-and-tested methods employed for centuries by carpenters and joiners.

DYNASTY OF IRONMASTERS

The third Abraham Darby came from a long line of Quaker ironmasters. When work began on the Iron Bridge, he was 26 years old and was following a family tradition of iron-craft begun by his grandfather Abraham Darby I – the first person to successfully smelt iron using coke – and continued by his father Abraham Darby II. The legacy of this dynasty of master iron forgers can still be seen throughout Coalbrookdale, the cradle of Britain's mining and metallurgical industries and a starting point for Industrial Revolution based on coal and steel.

A national treasure

The bridge at Coalbrookdale was opened to the public on 1 January, 1781. Initially, a toll was charged – the booths can still be seen. All was well until 1926, little more than two decades after the advent of the motor car, when despite additional bracing, cracks began to appear in the ironwork. The Iron Bridge was closed to road traffic eight years later. It immediately became part of Britain's industrial heritage and is now part of a UNESCO World Heritage Site.

Tolls for pedestrians continued until 1950, when ownership of the bridge passed to Shropshire County Council. It was restored in the 1970s, then starting in 1999 the cast-iron cladding was replaced by lighter steel plates. These renovations revealed that the smaller cast-iron parts had been cast in wooden moulds, while the larger elements were made by pouring molten metal directly into moulds created in piles of sand on the ground.

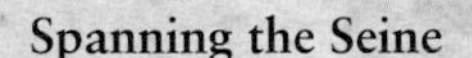

Spanning the Seine
The sweeping Léopold Sedar Senghor footbridge in Paris was designed by architect Marc Mimram and opened in 1999. It is named after the first president of independent Senegal.

METEOROLOGY – 1780

Forecasting the weather

The world's first meteorological society was founded in Germany in 1780. The idea of predicting the weather enthused scientists, but was slow to catch on in a Europe still reeling from the violent aftermath of the French Revolution. It took several decades and a terrible storm that hit the Crimean Peninsula during the war of 1853–6 to spark a widespread interest in meteorology.

Karl Theodor, Prince-Elector and Duke of Bavaria (1724–99) is hardly a household name, even among historians. Yet weather forecasting owes a great deal to this ruler of the Rhineland Palatinate, an historic realm in western Germany. For in 1780, he founded at Mannheim the world's first meteorological society. This scholarly body made the first serious attempt to set up a network of weather stations. Moreover, in order to standardise the various different scales and units of measurement in use at that time, the society supplied its weather monitors with identical measuring apparatus. It also established a system of symbols for recording atmospheric phenomena in a way that would be intelligible to everyone. The whole of Europe, with the predictable exception of Britain, stood poised to collaborate in this ambitious enterprise. But before it had a chance to get off the ground, the French Revolution broke out, and in the years that followed the Mannheim scheme fell victim to the political turmoil that engulfed the Continent.

CLASSIFYING CLOUDS

Cirrus, stratus, nimbus ... the magical-sounding terms for the various types of cloud were coined in 1802 by Luke Howard. This eccentric British chemist and amateur meteorologist devised a system for classifying clouds according to whether they formed a cluster (*cumulus* in Latin), a layer (*stratus*) or patterns like wispy curls of hair (*cirrus*), and whether they produced precipitation (*nimbus*; Latin for 'rain storm'). Using these basic categories, it was easy to define intermediate or compound types, such as cirrostratus or cumulonimbus. This highly practical nomenclature won out over a rival system proposed by the French biologist Jean-Baptiste de Lamarck. Howard's fame was also assured when the great German writer Goethe, a keen amateur scientist, dedicated a poem to him (*Howards Ehrengedächtnis* – 'In Honour of Mr Howard' 1817) and summarised his work in an essay four years later entitled *Howards Wolkenformen* ('Howard's Cloud Forms').

Cloudscape
A watercolour of cirrocumulus clouds by Luke Howard, who was made a member of the Royal Society in 1821.

Watchers of the skies
Charles II founded the Royal Observatory at Greenwich, a suburb on the eastern outskirts of London, in 1675 (inset, above). Its chief purpose was to make navigation on the world's high seas safer for the Royal Navy.

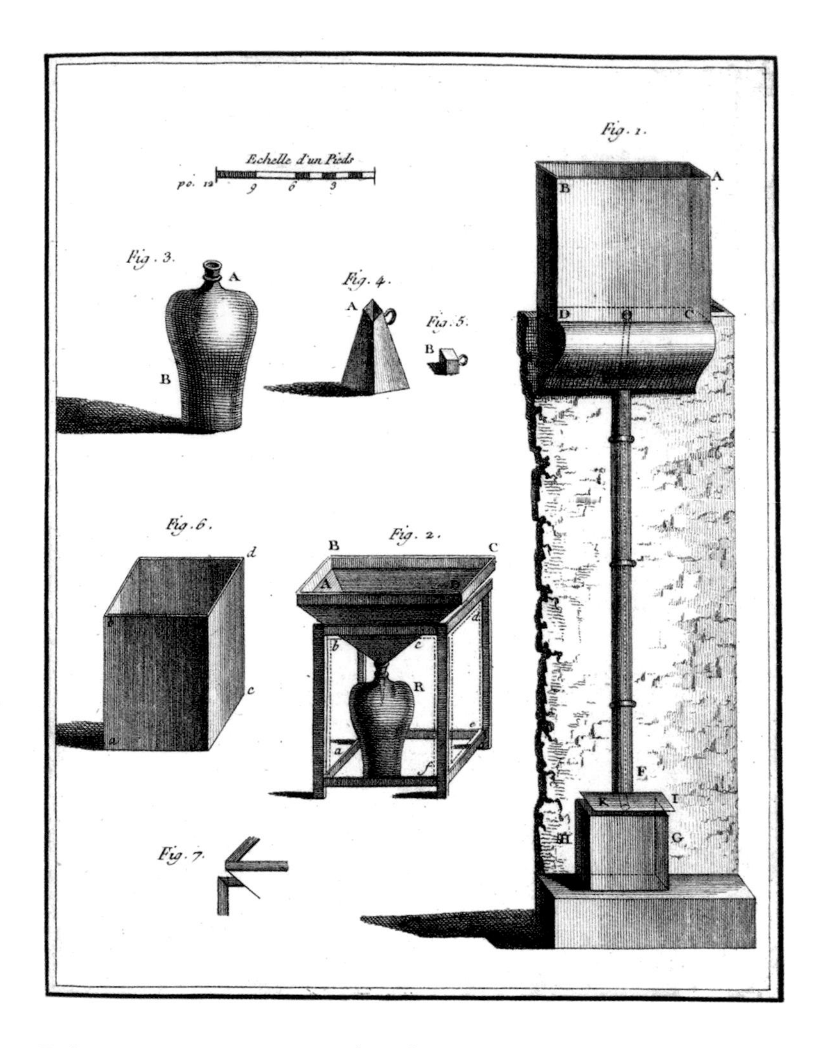

Rain gauges
An engraving from Père Louis Cotte's Memoirs of Meteorology *(1788), showing various types of rain-collecting devices.*

Wind force
A model of a rotating windmill anemometer (right) used at Canadian meteorological stations from 1875 onwards to measure the force of the wind.

Modest beginnings

This was a cruel blow to the infant science of meteorology. Thanks to various measuring instruments invented in the 15th century – the thermometer, barometer and anemometer – a great deal of weather data had already been collected right across Europe, and scientists had begun to investigate atmospheric phenomena. René Descartes, for example, had suggested to his fellow mathematician Marin Mersenne that they run an experiment between Paris, Clermont-Ferrand and Stockholm to see if they could forecast the weather using a barometer. Later, Edmund Halley charted the trade winds and concluded that atmospheric changes were caused by fluctuations in the radiated heat from the Sun as it warmed the Earth's surface. This was a key moment in the early history of meteorology, but it was still a long step from this to forecasting the weather.

In the early 19th century, scientists from different fields – including physicists François Arago, Heinrich Brandes and Henri Becquerel, naturalist Jean-Baptiste Lamarck and chemists Christophorus Buys Ballot and John Dalton – took a keen interest in the atmosphere. But meteorology still struggled to gain recognition as a science in its own right. All attempts thus far to forecast the weather had met with only limited success and people were sceptical about its usefulness. Even if farmers were forewarned about a hail storm, it would not help them to save their crops from damage.

Storm damage

It took a catastrophic weather event for people to sit up and take notice of meteorology. On 14 November, 1854, with the Crimean War at its height and the British and French fleets blockading the port of Sebastopol, a violent hurricane suddenly blew up, sinking four warships and scores

MEASURING RAIN AND WIND

People have measured rainfall since earliest times. Yet the first rain gauge bearing any resemblance to modern instruments was only created in 1722, by an Englishman, the Reverend Samuel Horsley. Horsley's gauge consisted of a large funnel attached to the top of a graduated glass cylinder. Rain gauges have hardly changed since; the most important progress in this field came in the late 19th century, with the introduction of a standard-diameter collection vessel, which made the collation of data much easier. A far trickier task, measuring wind strength, goes back to the Italian Renaissance and the polymath Leon Battista Alberti – painter, mathematician and rediscoverer of the technique of single-point perspective. Alberti invented a rudimentary version of the pressure-plate anemometer, a simple device that indicated wind strength through the angular deflection of a flat plate perpendicular to the wind. The English scientist Robert Hooke improved considerably on this technology in 1672, when he designed a rotating windmill anemometer, in which the speed of rotation was measured by gears that punched holes in a paper strip.

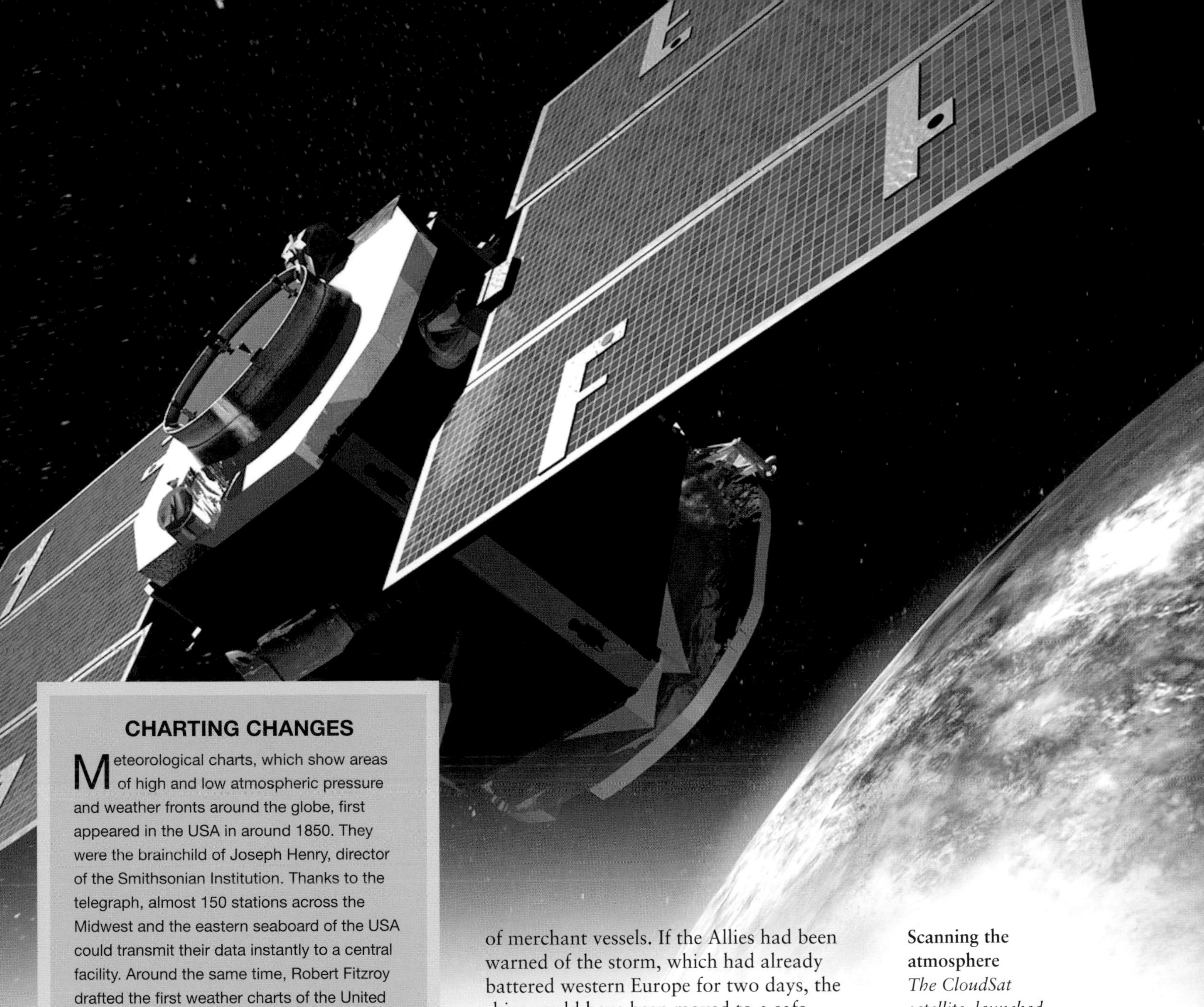

Scanning the atmosphere *The CloudSat satellite, launched by NASA in 2006, analyses the composition of clouds from outer space. Together with five other satellites, which orbit the Earth at intervals of just a few minutes, it is part of a joint Franco-American programme to gain a better understanding of the Earth's atmosphere and climatic events.*

CHARTING CHANGES

Meteorological charts, which show areas of high and low atmospheric pressure and weather fronts around the globe, first appeared in the USA in around 1850. They were the brainchild of Joseph Henry, director of the Smithsonian Institution. Thanks to the telegraph, almost 150 stations across the Midwest and the eastern seaboard of the USA could transmit their data instantly to a central facility. Around the same time, Robert Fitzroy drafted the first weather charts of the United Kingdom. Fitzroy's study of atmospheric changes day by day set the tone for modern weather forecasting. The first daily weather forecasts appeared in *The Times* in 1860.

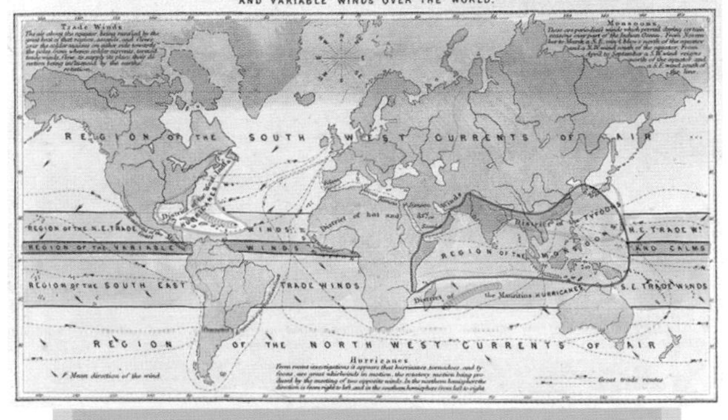

An early weather chart *produced in 1850 by the Smithsonian Meteorological Project.*

of merchant vessels. If the Allies had been warned of the storm, which had already battered western Europe for two days, the ships could have been moved to a safe anchorage. Urbain Le Verrier, director of the Paris Observatory, chaired an enquiry into the disaster in 1855 leading to the setting up of a chain of weather stations.

Meanwhile, the British Meteorological Society was founded in 1850, with balloonist James Glaisher as a founder member, to establish a 'general system of observation'. In 1854 Vice-Admiral Robert Fitzroy, a pioneer of meteorology who had also captained HMS *Beagle* during Charles Darwin's voyage, was put in charge of a new government department set up to deal with the collection of weather data at sea. Fitzroy arranged for ships' captains to provide information from calibrated instruments, including barometers of his own devising. Following the loss in a storm of the ship *Royal Charter* in 1859, Fitzroy developed prediction charts which he called 'forecasting the weather'. *The Times* newspaper published the first daily weather forecasts from 1860.

ARTIFICIAL INSEMINATION – 1780

Exploring the origins of life

The world's first successful artificial insemination was carried out on a dog by the Italian biologist Lazzaro Spallanzani. It proved conclusively that spermatozoa – which scientists had first observed under the microscope a century earlier – played a vital role in animal sexual reproduction. Yet Spallanzani never realised the full implications of his discovery.

Italian animalcules
A plate from Spallanzani's 1776 work, Tracts on the natural history of animals and vegetables *(opposite). Spallanzani made numerous references in the work to 'spermatic animalcules', his misleading term for sperm cells.*

At the end of the 18th century, how living organisms reproduced still remained a mystery. The spermatozoa and the ovum had been identified in the preceding century, but scientists had not yet fathomed the role played by either – or how an individual embryo was formed. There were a mass of competing and confusing explanations, mostly coloured by religious beliefs. One of the many researchers who set himself the task of clarifying this mystery was the eminent Italian biologist, Abbot Lazzaro Spallanzani. In 1780, he succeeded in artificially inseminating a spaniel bitch on heat, who subsequently bore a litter of puppies. This was the first procedure of its kind ever conducted and it marked the culmination of a long series of experiments on reproduction, which Spallanzani had begun some 15 years earlier.

Tireless researcher
A print from 1881 showing Lazzaro Spallanzani at work a century earlier, investigating digestion in birds. These studies highlighted the role that gastric juice played in digestion, proving that it was not just a mechanical process but also a chemical one.

Frogs in trousers

Spallanzani's experiments had begun with him slipping tiny, specially made impermeable leggings onto male frogs, in order to collect their semen and weigh it – a highly controversial procedure at the time. He chose frogs because fertilisation of their eggs takes place externally rather than inside the female's body. The female lays her eggs and the male simply covers them with his sperm, an arrangement that was ideal for the purposes of scientific observation. The outcome of Spallanzani's curious experiment was that the eggs did not develop into tadpoles, so confirming his hypothesis that sperm was vital in the conception of a living organism.

AN INQUIRING MIND

Lazzaro Spallanzani's research in the life sciences did not stop at pioneering work on fertilisation. He also studied the flight of bats, autotomy (the regeneration of limbs by certain lizard species), animal digestion and how the circulatory system works. His investigations prefigured much of the groundbreaking work in experimental medicine that was to come in the mid to late 19th century. In addition, his extensive travels around the Mediterranean kindled his interest in other scientific fields. He studied Vesuvius and other volcanoes, on the Aeolian Islands and Sicily, which furnished him with material for a work on vulcanology. He also investigated meteorology, and copper and iron mining in Turkey.

PRECURSOR OF PASTEUR

In the 1740s, John Turberville Needham, a Scottish naturalist and clergyman, did some experiments that involved boiling gravy in closed containers, then letting them stand for a while. On opening the jars, he discovered that micro-organisms had begun to develop on the surface of what was non-living material. For Needham, the reasons for this were clear: life forms arose through spontaneous generation. Spallanzani repeated these experiments between 1760 and 1771, taking great care to seal the containers – and obtained the opposite results: no organisms grew in the jars so long as they were hermetically sealed. This proved that microbes come from the surrounding air, and that boiling a liquid sufficiently vigorously kills them. This opened the way for Louis Pasteur's later breakthroughs in germs and sterilisation.

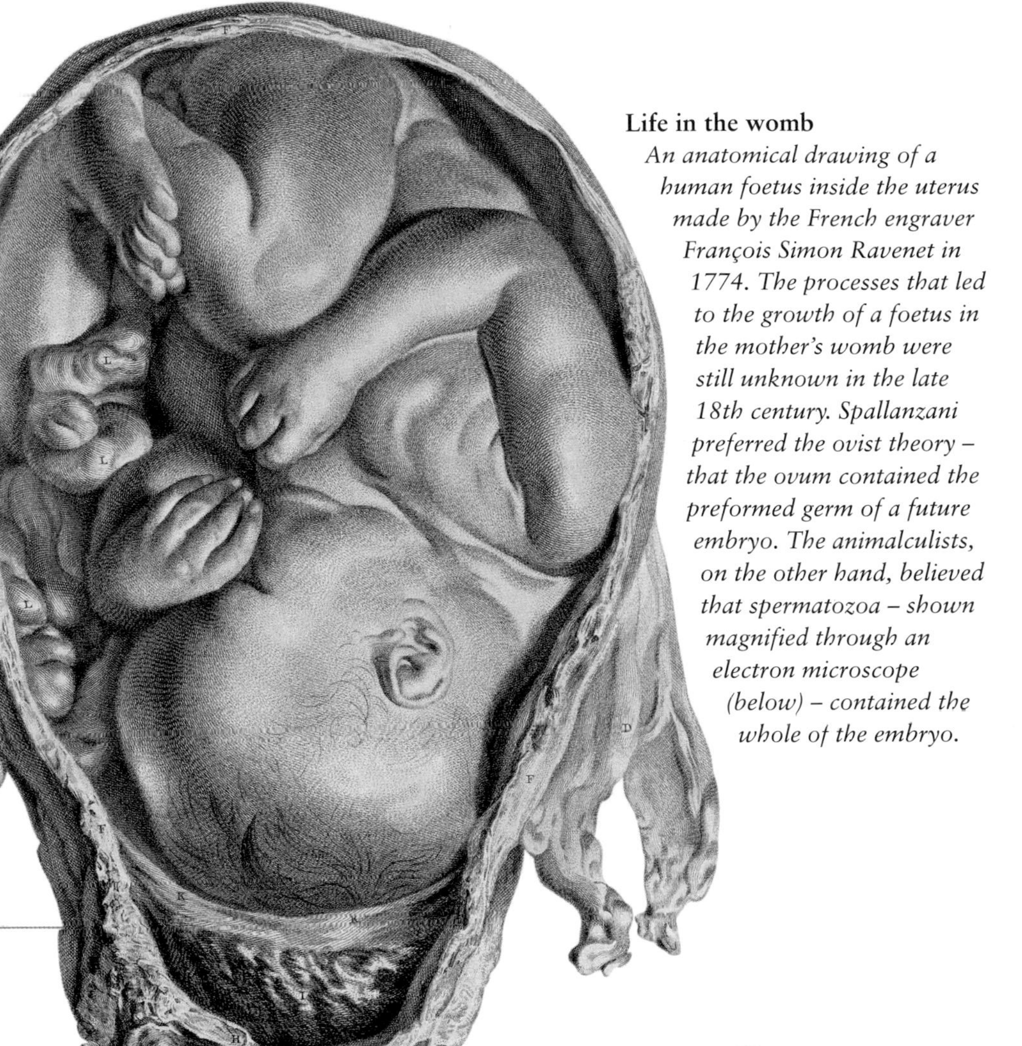

Life in the womb
An anatomical drawing of a human foetus inside the uterus made by the French engraver François Simon Ravenet in 1774. The processes that led to the growth of a foetus in the mother's womb were still unknown in the late 18th century. Spallanzani preferred the ovist theory – that the ovum contained the preformed germ of a future embryo. The animalculists, on the other hand, believed that spermatozoa – shown magnified through an electron microscope (below) – contained the whole of the embryo.

Spallanzani passed the sperm he had collected from the frogs through a series of ever finer filters, showing that if a very close-meshed sieve was used, fertilisation did not take place. Yet for all his innovation, he remained a convinced ovist, that is, he believed that the female reproductive cell contained the whole embryo of the future organism and that the sperm merely 'awakened' it by touch. He also thought that spermatozoa were parasites, and consequently failed to identify the key role that they played in reproduction.

A commonplace procedure

Many decades were to pass before biologists finally realised that the spermatozöon and the ovum combine in the reproductive process. Eventually it was demonstrated conclusively that fusion between sperm and egg was an essential prerequisite for fertilisation in sexually reproducing organisms. As a result, artificial insemination of certain species, such as cows and horses, came on in leaps and bounds around the turn of the 20th century as a way of improving animal husbandry. Thereafter, it was used to combat infertility in humans. Meanwhile, embryology became a recognised medical discipline, so lifting the curtain on a new realm of scientific enquiry – genetics.

The watch winder *c*1780

Historic timepiece
The No.15 sel-winding watch invented by Abraham Louis Breguet in about 1780 (left). Known as the 'Perpetuelle', it was also fitted with an anti-shock mechanism to protect it from knocks and jolts when carried in a pocket. Breguet made timepieces for Louis XVI and Marie Antoinette, and he offered this watch as a gift to the Duke of Orléans.

Prior to the 1780s, the only way of winding a watch was to open it up and tighten up the spring with a key. But in around 1780 Abraham Louis Breguet, a Swiss clockmaker living in Versailles, devised a mechanism whereby a watch could be rewound using a button mounted on the edge of the case. This elegant and simple solution was fitted to all watches, carriage clocks and public clocks, and remained the standard mechanism for rewinding watches until the advent of quartz mechanisms in the late 20th century.

The goniometer 1782

Precision instrument
A 19th-century horizontal-circle goniometer (right). Modern instruments that work on the principle of refracted X-rays can now show crystallographers the minutest detail, right down to the arrangement of atoms within a material.

The measurement of angles became a scientific discipline in the late 1700s, thanks to the emerging field of crystallography. We now know that the angle formed by two adjacent faces of any given crystal is constant. Indeed, this is the principal criterion for identifying a crystal. But originally the only way to determine this angle was to fold a piece of card around the crystal and measure the resulting angle with a protractor.

In 1782, French mineralogist Arnould Carangeot devised the contact goniometer, a semicircular protractor with two metal arms pivoted together that could be applied simultaneously to two adjacent faces of a crystal – the faces had to be absolutely flat to allow the arms to make proper contact. Each arm comprised an alidade (a sighting device) and the interfacial angle was read off a graduated scale marked on the protractor. The goniometer was first in a line of instruments for measuring angles which, over time, found applications in many different areas.

The power of light

In 1809 the British chemist William Hyde Wollaston developed the reflecting goniometer, which utilises the reflective power of crystals. In this apparatus, the crystal is mounted onto a graduated circle. As daylight strikes one face of the crystal, it produces a beam of light. This reflected ray is aligned with a fixed point of reference – such as the edge of the table on which the instrument is resting. The operator then rotates the crystal about an axis parallel to the edge between two crystal faces, and brings the light ray reflected from a second face into the same position as that formerly occupied by the reflected beam from the first face. The interfacial angle is obtained by a simple sum, subtracting the angle through which the crystal had been rotated (indicated on the graduated circle) from 180°.

THE VERSATILE GONIOMETER

All sorts of instruments go by the name of goniometer, which comes from Greek gonia (meaning angle) and metron (meaning measurement). The common denominator is that they all measure angles, but they are used in a wide range of different fields, including surface science, navigation, engineering and even physical therapy.

Uranus 1781

Before 1781, the only planets known to humanity were the five that could be seen with the naked eye: Jupiter, Saturn, Mars, Venus and Mercury. But on 13 March of that year, William Herschel, a British astronomer of German extraction, noticed a 'curious nebulous star or perhaps a comet' beyond Saturn. Uranus had actually been sighted 17 times already – the first in 1690 by John Flamsteed – but due to the lack of sufficiently powerful optical instruments, it was mistaken for a fixed star.

Herschel was well equipped to study this new celestial body. He constructed his own telescopes and the one that he was using when he observed Uranus was 158 millimetres in diameter, with a focal length of 2.1 metres and a magnification of x227. Herschel soon ascertained that this was not a star like the thousands of others in the sky, because its outline was so crisp and clear. Having no tail, it could not be a comet. The clarity of the images Herschel saw through his telescope enabled him to measure the movement of this celestial body, and to reach the inescapable conclusion that it was a planet.

FACTS AND FIGURES

In order of distance from the Sun, Uranus is the third 'gas giant', after Jupiter and Saturn. To date, 13 rings have been counted around the planet; the last two were found by the HST (Hubble Space Telescope) in 2005.

Painstaking observation

Herschel spent a year taking further sightings to confirm his hypothesis. He also worked out that the distance between the Sun and Uranus was 19 astronomical units (the unit was based on the distance between the Earth and the Sun), equal to 2.9 billion kilometres. The scientific world was agog at this new discovery, and Herschel became something of a celebrity.

After Herschel's death, Uranus continued to deliver new surprises: discrepancies in its orbit seemed to suggest that a nearby, as yet undiscovered planet was exercising a gravitational pull on it. And indeed, some 60 years after the discovery of Uranus, research into its orbit by the French astronomer Le Verrier revealed the presence of the next planet in the Solar System – Neptune.

The great 40-foot telescope
The largest and most famous of Herschel's optical instruments was a reflecting telescope with a 12m focal length and an aperture 1.26m in diameter, which he built at his home, Observatory House in Slough, between 1785 and 1789. He had discovered Uranus with a much smaller telescope.

New clarity
The Hubble Space Telescope (HST) has given astronomers new insights into the Solar System and deep space. This image of Uranus shows its atmosphere (made of helium and hydrogen) in blue, its high-level clouds (orange and pink) and its rings.

HOMAGE TO A KING

Ever the loyal subject, Herschel proposed naming the new planet Georgium Sidus ('George's Star') in honour of George III. In turn, the King recognised his German compatriot's achievements by appointing him as his personal astronomer. But the name never caught on outside Britain. It was another German astronomer, Johann Bode, who ultimately dubbed the planet Uranus, after the Greek god of the sky, Ouranos. Herschel subsequently discovered two moons of Uranus, which he called Oberon and Titania, after two Shakespearean characters.

Soaring to the skies

In Greek mythology, Daedulus, master-builder of the labyrinthine palace of Knossos on Crete, fashioned wings for himself and his son Icarus to escape the island. But Icarus flew too close to the Sun, and as the wax on his wings melted, he fell. By the late 1700s, the Montgolfier brothers were about to turn the stuff of myth into reality.

First draft
The Montgolfier brothers' original design for a hot-air balloon, made in 1782 (right).

The remarkable story of ballooning began one cold night in November 1782, on the second floor of number 18, rue Saint-Étienne in the southern French city of Avignon, where Joseph de Montgolfier was seated in front of a roaring fire. Glancing up from a report that he was reading on the long French siege of the British stronghold of Gibraltar, ongoing since 1779, he was struck by the sight of his nightshirt, which he had hung above the fireplace to warm. It was buttoned up to the neck and the hot air rising from the fire made it billow out. A brilliant idea struck him: could the Franco-Spanish forces simply sail over the British defences in balloons to break the siege?

Joseph called for thread, needles, scissors and pieces of silk taffeta, and set about cutting out pieces of cloth. He stitched these into a polyhedron shape, firmly sewn together on five of its faces, with the sixth left open. In a state of mounting excitement, he held his crude fabric envelope above the fireplace – and let it go. He was thrilled to see it waft right up to the ceiling. There and then, he scribbled a note to his younger brother Étienne: 'Get in a supply of cordage, quick, and you will see one of the most astonishing sights in the world.'

From the sticks to the royal court

Keen amateur scientists, the Montgolfier brothers had read the works of the British chemist Joseph Priestley, in which he described various different kinds of air. They realised that it was 'rarefied air', which Priestley said was generated by fire, that had kept Joseph's rudimentary balloon aloft. The Montgolfier family owned a large paper mill at Vidalon-lès-Annonay in the remote Ardèche region of southern France, and there they began experimenting with different balloons made from canvas or paper. On 14 December, 1782, the first hot-air balloon worthy of the name – a paper globe with a volume of 3 cubic metres – rose into the skies above the factory garden.

This was a promising start, but the Montgolfiers had bigger things in mind. They were convinced that balloons would enable people to 'send signals to the ground, convey messages into besieged cities, and conduct experiments on the electricity in clouds'. They constructed a large balloon made from sackcloth, stiffened inside with a triple thickness of paper and held together with stout rope. Propulsion was provided by a brazier fitted beneath the mouth

THE ARCHIMEDES' PRINCIPLE

A law formulated by the ancient Greek scientist Archimedes states that 'any object, wholly or partly immersed in a fluid, is buoyed up by a force equal to the weight of the fluid displaced by the object'. The same law applies to any lighter-than-air gas subjected to Earth's gravity. When a balloon is filled with hot air, hydrogen or helium, the lift deriving from the force described by Archimedes offsets the weight of the balloon and causes it to rise. This was demonstrated for King John V of Portugal in 1709 by the Brazilian Jesuit missionary Bartolomeu Lourenço de Gusmão with his airship the *Passarola*, a gondola-like structure with a large sail canopy (see above).

Up, up and away
A contemporary print of the Montgolfiers' first unmanned balloon flight at Annonay on 4 June, 1783.

PROLIFIC INVENTORS

Though they will always be remembered for the hot-air balloon, the Montgolfier brothers came up with many other innovations. In 1777, Étienne produced the first woven vellum paper in France. Joseph is credited, among other things, with inventing the automatic hydraulic ram, which enabled water to be pumped to a height of several metres using water pressure alone. Joseph was appointed a director of the College of Arts and Crafts and in 1807 was made a member of the French Academy of Sciences, 11 years after his brother.

of the balloon and fuelled by burning straw and lambswool. It lifted off from the main square in Annonay on 4 June, 1783, and made a flight lasting around 12 minutes. It came down to earth 2.5 kilometres away, where it promptly, and predictably, caught fire.

The *Mercure de France* reported the event on 26 July, and Louis XVI demanded that he be given a demonstration. For this grand occasion, the Montgolfiers prepared a large balloon some 13 metres in diameter, made by stitching together 900 metres of blue serge cotton. Emblazoned with the royal 'XVI' and christened *Le Reveillon*, this balloon rose into the air on 19 September, 1783, from a courtyard at the Palace of Versailles. An excited crowd of people watched it go, its open gondola carrying a sheep, a rooster and a duck. The balloon touched down safely at Vaucresson, 3.5 kilometres away, with the animal crew alive and well. The *Reveillon* had ascended to a height of 2,000 metres during its brief flight. Étienne later wrote to his wife with this whimsical account of the occasion: '19 September, 1783: the news from on board the hot-air balloon *Reveillon* is that the crew are all doing well, and having arrived safely at their destination, now have quite an appetite. And that's really all there is to report about our three intrepid aeronauts – given that they don't know how to write and that nobody has bothered to teach them French. So one of them has nothing to say but "quack, quack", the second "cock-a-doodle-doo", while the third's sole response to questions is "baa"!' At least the sheep, who was jokingly christened Montauciel ('ascend to the skies'), was rewarded for his exploits by living the rest of his life in the Royal menagerie.

As for the Montgolfier brothers, they were knighted by the King, who bestowed on them the motto *Sic itur ad astra,* 'thus we shall get to the stars', a quotation from the Roman poet Virgil. Exhilarated by their successful venture, they felt that the time was ripe for people to experience the thrill of flight.

Flight of fancy
Balloon mania soon swept Europe. This humorous lithograph of a fashionable lady borne aloft by her ballooned-out skirt was produced in 1783.

BALLOON MANIA

The first balloon flight was widely reported throughout Europe, and soon fashionable society was in the grip of a balloon mania. The craze for ballooning even found its way into the literature, song and dance of the period, while traders jumped on the bandwagon, naming all sorts of things after the pioneering Montgolfier brothers – hairstyles, hats, dresses, fans, crockery, snuff boxes, waistcoats, walking sticks. 'Everything in Paris is balloon-shaped', wrote the journalist Antoine de Rivarol.

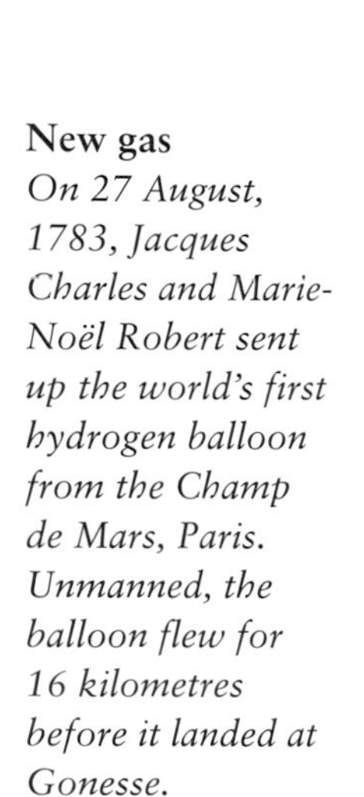

New gas
On 27 August, 1783, Jacques Charles and Marie-Noël Robert sent up the world's first hydrogen balloon from the Champ de Mars, Paris. Unmanned, the balloon flew for 16 kilometres before it landed at Gonesse.

The first aeronauts

The first person to ascend in a hot-air balloon was François Pilâtre de Rozier. Born in Metz in 1754, he had moved to Paris aged 18 to seek fame and fortune, but ended up teaching chemistry at the academy in Reims. He was fascinated by the Montgolfiers' experiments and assisted at the September launch. On 15 October, 1783, he made an ascent in a tethered balloon, rising 30 metres above the garden of a wallpaper manufacturer on the rue de Montreuil in Paris. Then, on 21 November that same year, he and the Marquis d'Arlandes made the first manned balloon flight, from the lawn of the Château de la Muette on the edge of the Bois de Boulogne. The venture was perilous – the pair ignored an instruction from Louis XVI that they use two condemned men instead. With a command to 'Release all guy ropes!', the two men began their ascent in a balloon with an envelope of 2,000 cubic metres capacity and a straw-fuelled brazier.

The assembled citizens of Paris, including the great American scientist Benjamin Franklin, craned their necks to watch these first aeronauts rise some 1,000 metres into the air. They drifted over the Seine, passed between the Military Academy and Les Invalides, then finally set the craft down safely at the Butte-aux-Cailles on the outskirts of the city. They had covered 9 kilometres in 25 minutes. Their pioneering flight ushered in a new age where people would no longer be constrained by the bonds of Earth.

The next manned balloon flight took place on 1 December that same year, when physicist, mathematician and chemist Professor Jacques Charles, accompanied by the instrument maker Marie-Noël Robert, took off in the Globe from the Tuileries Gardens watched by a crowd estimated at 400,000, an enormous

A NEAR DISASTER

Joseph Montgolfier's first (and last) balloon flight took place above Lyons on 19 January, 1784. He was accompanied by six others – Pilâtre de Rozier, Prince Charles-Joseph de Ligne from the Austrian Netherlands (modern Belgium), the Count of Laurencin, who was funding the venture, the Count of La Porte d'Anglefort et d'Antraigue, the Marquis of Dompierre and a man named Fontaine, who had built the great balloon to carry all these VIPs. Named the *Flesselles*, it was one of the largest balloons ever built, measuring 23,270 cubic metres. Some 100,000 people turned out to watch the launch. Twelve minutes later, it plummeted back down to earth. Buffeted by winds that tore a huge gash in the fabric of the balloon, it suffered a sudden and catastrophic loss of height. Joseph Montgolfier sustained three broken teeth, while the prince sprained his ankle.

gathering for the period. Before they set off, Charles is reputed to have asked Étienne de Montgolfier, who had come along simply out of curiosity, to release a small balloon that would show the prevailing wind direction, with the following words: 'To you, Sir, falls the honour of showing us the route skyward!'

Rival gases

Charles and Robert stayed aloft for almost two hours before finally alighting in a meadow in Nesles-la-Vallée north of Paris. Yet no sooner had the *Globe* set down than Charles took to the air again, climbing to an altitude of almost 4,000 metres in 10 minutes. The rapid ascent caused him some discomfort; he complained of pains in his ears that he ascribed to 'the expansion of gases contained within the cell tissue of the human body, as well as the cold atmosphere'. The balloon, which was made of red and yellow striped silk, was the first hydrogen-filled balloon capable of carrying human beings. Hydrogen was 14 times lighter than air. It had been discovered by the British chemist Henry Cavendish in 1766. Known at the time as 'inflammable air', it was obtained by dissolving iron filings in vitriol (sulphuric acid), but suffered from one major handicap, as its old name indicated: it was highly flammable, which made it extremely dangerous to work with. A long-running

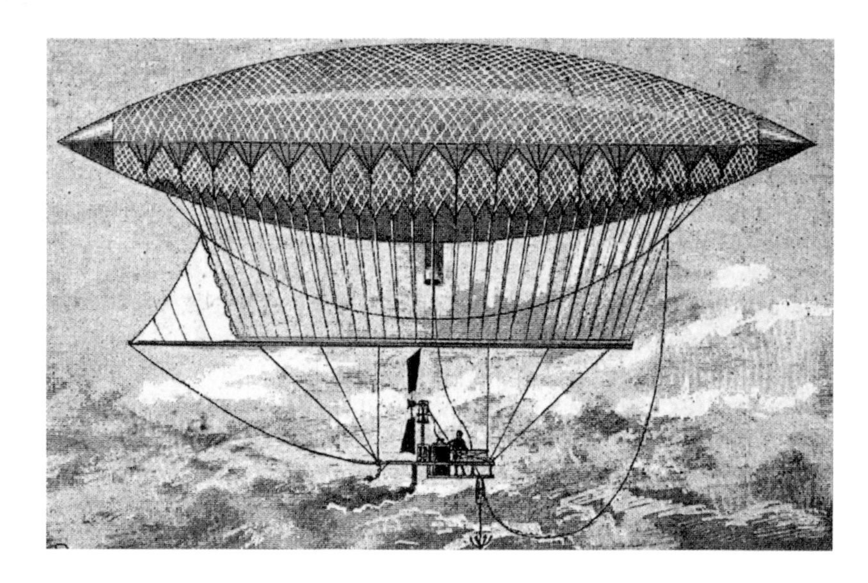

Prototype airship
The world's first dirigible airship (above), built by Henri Giffard, made its maiden flight on 24 September, 1852.

SPY IN THE SKY

The French military soon realised the balloon's potential for observing enemy movements in wartime. By the end of the Directory era (1795–9), the country's military balloon fleet already numbered 23 hydrogen-filled craft with suitably warlike names, such as *Le Martial, L'Hercule* and *L'Entreprenant*. But when Napoleon came to power, he was unimpressed by the time it took to inflate the balloons (12 hours) and the complex equipment needed to maintain and launch them, and so he disbanded the military balloon units. The next time they were used in a conflict was 1870-1, during the Franco-Prussian War.

Over the sea and under
It did not take long for the balloon to be deployed in warfare. Napoleon had a plan to invade England with a fleet of 100 huge hot-air balloons, each carrying up to 100 men. This fanciful engraving shows this and another ambitious, unfulfilled invasion plan – a Channel tunnel.

rivalry ensued in ballooning circles between 'Montgolfierists' – exponents of hot air – and 'Charlists', who favoured the use of hydrogen.

Whatever the lifting agent, the late 18th century saw a great proliferation of balloon flights. On 4 October, 1784, James Sadler became the first Englishman to make a balloon ascent, from Christ Church Meadow, Oxford. The following year Jean-Pierre Blanchard, inventor of the parachute, and the American John Jeffries made the first air-borne crossing of the English Channel, from Dover Castle to Guînes, near Calais, in a hydrogen balloon fitted with steering oars and a propeller.

Another milestone came on 24 September, 1852, when the engineer Henri Giffard flew 27 kilometres, from the Paris Hippodrome to Trappes on the western outskirts of the city, in a 300-cubic-metre capacity hydrogen balloon. What was revolutionary about Giffard's craft was that it was both dirigible (steerable) and powered by a small steam engine, the first balloon to incorporate such innovations. In September 1862 the English balloonists James Glaisher and Henry Coxwell made the first of several ascents together. Glaisher took meteorological instruments with him and took frequent readings until he lost consciousness at 29,000 feet. Coxwell claimed that he reached 36,000 feet before descending – this was probably an exaggeration, but they undoubtedly set a record height for a balloon ascent. Other balloon exploits of the late 19th century included a daring aerial escape from Paris by the French politician (later prime minister) Léon Gambetta in October 1870, during the Franco-Prussian War.

Golden age of the dirigible

In the early years of the 20th century, popular attention turned increasingly to the aeroplane. Yet the development of light and powerful internal combustion engines also fostered the growth of the airship. In Germany, more than 40,000 passengers were carried on board Zeppelins between 1900 and 1914. But it was the ensuing interwar period that became the golden age of the dirigible. In their heyday, more than 1,300 were in service worldwide, making regular Transatlantic flights and carrying polar explorers on daring expeditions.

Germany, the USA, France, Italy and Britain all treated airship construction as a matter of national prestige and vied with one another to build ever larger, more luxurious machines. Unfortunately, these elegant craft were prone to catastrophic accidents. In 1930, the British airship R101 crashed in northern

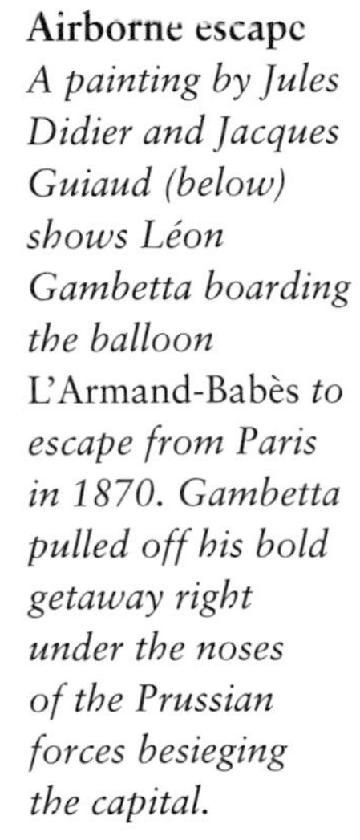

Airborne escape
A painting by Jules Didier and Jacques Guiaud (below) shows Léon Gambetta boarding the balloon L'Armand-Babès *to escape from Paris in 1870. Gambetta pulled off his bold getaway right under the noses of the Prussian forces besieging the capital.*

Pride of the Reich
The Nazis used the graceful and lavishly appointed airship Hindenburg *as a propaganda tool to show off the technological prowess of Hitler's Germany; the luxurious on-board restaurant is shown below. Her giant gasbags were originally designed to be filled with helium, but after Hitler came to power in 1933 the US government placed an embargo on the sale of this inert gas – on which it had a virtual monopoly – to Germany. As a result, the* Hindenburg's *designers were forced to used hydrogen. On 6 May, 1937, the airship was engulfed in flames and completely destroyed while trying to land at Lakehurst, New Jersey; 36 people lost their lives in the disaster.*

Freedom of the skies
Today, hot-air ballooning is primarily a leisure activity. In many parts of the world – as here (below) above the heavily wooded countryside of Quebec, Canada – balloon companies take paying passengers on pleasure flights, giving them stunning bird's eye views of natural spectacles and cities.

France just hours after embarking on a test flight to India, killing 48 passengers and crew. Seven years later, on 6 May, 1937, the German airship D-LZ129 *Hindenburg*, the largest and most graceful hydrogen-filled airship ever built, burst into flames while coming in to land at Lakehurst, New Jersey, after a Transatlantic flight. The loss of the *Hindenburg* and 36 lives spelled the end for passenger airships, which only resurfaced again on a modest scale in the 1970s.

Airships gained a new lease of life when the inert gas helium replaced hydrogen as a lifting agent. Dirigibles, also known as 'blimps', are now mainly used as flying billboards or for tasks where slowness and stability are an advantage, such as monitoring the weather, aerial surveying or coastguard duties. Faster than ships and less polluting than aircraft, airships are even now being reconsidered as a viable way of carrying freight worldwide. By 2040, as fuel stocks dwindle, the skies may be filled with thousands of airship freighters.

STEAMSHIPS – 1783

The bell tolls for sail

The advent of the steam engine as a method of propulsion for ships would turn the maritime world on its head, but the transition from sail to steam was by no means swift or trouble-free. The new mode of transport was first adopted in the United States, and by the end of the 19th century it was firmly established in Europe as well.

American pioneer *The very first steamboat in the USA was a 14m-long craft built by John Fitch, who tested it on the Delaware River in 1787. Its top speed was 6mph against the current.*

On 15 July, 1783, a large crowd gathered on the banks of the River Saône, near Lyons. They had come to see an attempt at steam navigation on water by the Marquis Claude François de Jouffroy d'Abbans, using an experimental vessel he had constructed named the *Pyroscaphe*. This substantial clinker-built wooden boat – measuring 45 metres in length and 5 metres in the beam – was equipped with two steam-driven, side-mounted paddlewheels. The aim of the experiment was to see whether a boat could be propelled along independently of the winds and currents.

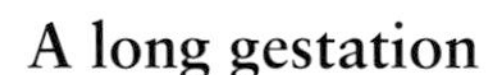

A long gestation

The Marquis' demonstration was the latest in a long line of endeavours to find a means of propulsion on water other than oars or sails. Sometime between the 3rd and the 1st centuries BC, the Romans had experimented with paddlewheels driven by human muscle power. The idea was reprised in China in around AD 1200, then in 16th-century Europe, by both Leonardo da Vinci and the Spanish inventor Blasco de Garay.

In *c*1687 the French engineer Denis Papin designed a revolutionary engine in which the condensation of steam in a cylinder drove a piston. In 1707, he mounted a version of this engine on a prototype sidewheel steamboat on the River Weser, but the vessel was destroyed by a mob of worried, angry boatmen. In 1736 an Englishman named Jonathan Hulls fitted out a boat with an 'atmospheric engine' – a type of steam engine invented 24 years earlier by Thomas Newcomen and Thomas Savery to pump water from a mine. The final significant step before Jouffroy d'Abbans' breakthrough was in 1775, when the French inventor Jacques-Constantin Périer tried unsuccessfully to run a steamboat on the River Seine – the engine lacked sufficient power to move the vessel.

Making history *The trial steamboat run of Jouffroy d'Abbans on the River Saône in 1783 was watched by a cheering crowd of thousands. It was the first truly successful test of a boat moving under its own power.*

AN INTERIM TECHNOLOGY

The paddlewheel, which was first used in watermills, was one of the first forms of mechanical propulsion to be applied to boats. Paddlewheels were driven by pressurised steam from boilers fired by wood or coal. Mounted either on a vessel's stern or more commonly amidships, paddlewheels usually comprised 10 or more radial boards and were reversible to enable the ship to proceed ahead or astern. But paddlewheels were fragile, unwieldy and relatively inefficient, and were eventually replaced by the screw propeller.

A sad end

Jouffroy d'Abbans' first attempt to launch a steamboat was in 1778, at the little town of Daume-les-Dames on the banks of the River Doubs in eastern France. The boat was fitted with a small version of a Newcomen steam engine and complex paddles designed to work like the webbed feet of a duck. As the paddles entered the water, wooden shutters were supposed to open like hand-held fans, make the stroke, then close again to allow the boat to move smoothly forward. Unfortunately, in strong currents the shutters failed to open.

Some time before his River Saône trial, the inventor swapped the webbed paddles for sidewheels. As he fired up the *Pyroscaphe*'s engine, a huge plume of smoke billowed from her smokestack, then the vessel began to move. She chugged steadily upstream for a quarter of an hour. Yet for all the undoubted success of the trial, the French Academy of Sciences failed to grasp its importance and refused Jouffroy d'Abbans permission to demonstrate his boat in Paris. The Marquis eventually went bankrupt and fled the country during the Revolution. He returned to France under Napoleon and resumed his research, but received no official funding and made little headway. By the time he died – of cholera in Paris in 1832 – steam navigation was still struggling to hold its own against the railway, which Louis Philippe I promoted and subsidised. Meanwhile, other engineers had already picked up the baton.

America takes up the challenge

In 1787 the American inventor James Rumsey succeeded in travelling some 4 miles up the Potomac River on a boat powered by a hydraulic jet propulsion system first proposed by the Swiss mathematician Daniel Bernoulli. A large pump driven by a steam engine squirted a jet of water from the stern of the vessel, thus propelling the boat forwards. Meanwhile, in Scotland, William Symington produced an improved version of James Watt's steam engine and installed it in a tugboat named the *Charlotte Dundas*. During a trial on the Forth and Clyde Canal near Glasgow, in March 1803, she towed two 70-ton lighters against the wind for some 20 miles. But it was the American inventor Robert Fulton who made steam navigation a viable form of transport.

Travelling to France in 1797 to study steam propulsion, the 34-year-old Fulton was given great encouragement and support by the newly appointed US ambassador there, Robert R Livingston, one of the original signatories of the US Declaration of Independence. Fulton's first prototype boat was a spectacular failure, breaking her back under her own weight and sinking just before her maiden voyage. Undeterred, Fulton and Livingston built a second boat, a 20-metre paddlewheel steamer, and prepared to demonstrate it on the River Seine on 9 August, 1803. Casting off from Chaillot, Fulton's craft travelled up the river at the speed of a fast-walking pedestrian, executed various

Seine steamer
A model of the experimental steamboat that Robert Fulton demonstrated in Paris in 1803 (below, top). It had an 8-horsepower steam engine and paddlewheels almost 4 metres in diameter. With the current, it reportedly reached speeds of over 6mph.

Not such a folly
At first, the public were sceptical of steamships. 'Fulton's Folly' was the unflattering nickname given to Robert Fulton's Claremont *(bottom), which plied the Hudson River from New York to Albany.*

manoeuvres, and even took on board several members of the Academy. Fulton hoped to persuade the French government to take an interest, but Napoleon was unimpressed.

Yet Fulton's trials had shown that a small crew could navigate a steamboat long distances regardless of the weather. He continued his work on his return to the United States and in 1807 inaugurated a regular passenger steamboat service up the Hudson River from New York to Albany, a distance of some 120 miles. The *Claremont*, a 40-metre long, flat-bottomed steamer, made the trip in around 30 hours at an average speed of 4.5 knots. Many of the country's waterways were ideally suited to steam navigation, and steamboats soon became a common sight on the Delaware, Mississippi and Missouri rivers. They carried not only passengers but also cotton from the slave plantations of the South to the textile mills of the North.

Steam's long route to supremacy

In Europe too, steam navigation was starting to be accepted. From 1812 onwards, the paddle steamer *Comet* ran a regular passenger service on the Clyde between Glasgow and Greenock. Four years later, the *Elise* made the first steam crossing of the Channel. Yet on longer voyages, steamships were no match for the clippers – the fast, graceful, multi-masted sailing ships that were used to rush valuable and perishable cargoes to Europe and America. Although steamships were in no danger of being becalmed, they were too fragile and temperamental to brave the great oceans of the world. Large steamers also required an army of stokers working non-stop, while the requisite coal took up so much space on board that there was scarcely room left for passengers or cargo. And so, for their first few decades, steamships remained little more than curiosities, exciting as much fear and derision as they did admiration.

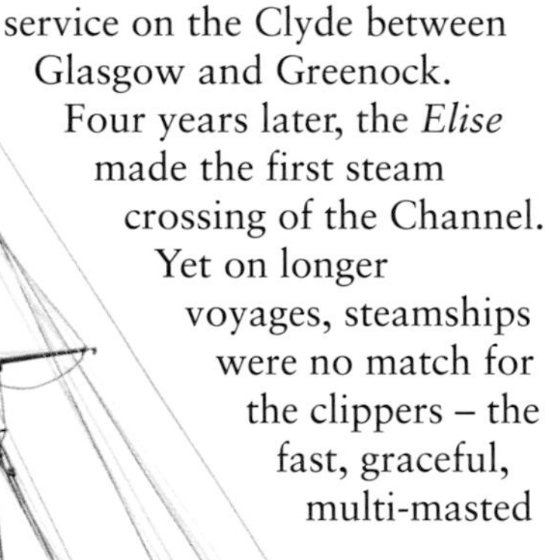

Hybrid ship
A model of the steam-and-sailing ship Savannah, *the first vessel to use steam on a Transatlantic voyage. Her smokestack could be rotated to suit the prevailing wind direction, to prevent sparks from her boilers setting the sails alight.*

Crossing the pond
The Great Western *leaving Bristol bound for New York on 8 April, 1838. Launched the previous year, this passenger vessel was owned by the Great Western Steamship Company, a subsidiary of the Great Western Railway enterprise and intended to be an extension of that service. She was the first of three ever larger and more technologically advanced ships designed by the Victorian engineer Isambard Kingdom Brunel (1806–59). Brunel also built the Clifton Suspension Bridge in Bristol.*

A TALE OF TWO STEAMERS

On 23 April, 1838, two ships entered New York harbour just a few hours apart. Both were greeted with great jubilation. They had crossed the Atlantic powered solely by steam – the first vessels ever to do so. The *Sirius* left Cork in Ireland on 6 April with 40 passengers on board. Generally, she had a smooth, incident-free passage, but shortly before reaching America she ran out of fuel. Determined to complete the crossing, her captain ordered that all the ship's wood fittings – including furniture, panelling and spars – be burned to keep her going. Meanwhile, Brunel's *Great Western* had left Bristol on 8 April. She was almost twice the size of the *Sirius*, and sailed with a complement of 150 on board. She was faster than her rival, averaging 9 knots against 6.7 for the *Sirius*, but she just failed to catch the competition and had to be content with being the second steamship to cross the Atlantic.

The first transatlantic steamers

In June 1819, the journey of the American sail-steamer *Savannah* from Georgia in the USA to Liverpool passed almost without comment. She had a steam engine driving two large sidewheel paddles, but given the limited space for coal in her holds, she was also rigged out with three masts. In all, the *Savannah* was under steam for less than 4 of the 27.5 days that it took her to complete the crossing. Otherwise, she was a conventional sailing ship.

In contrast, an enthusiastic crowd cheered the arrival of the small steamship *Sirius* in New York on 23 April, 1838. *Sirius* had just crossed the Atlantic in 17 days on steam power alone. By the following year, Britain had three regular transatlantic steamship services, with crossings averaging 15–17 days, as compared to 22–34 days for sailing ships.

During the War of 1812, Robert Fulton had built the world's first steam-powered warship, *Demologos*, a floating battery designed to defend New York against British attack; hostilities ceased before she saw action. Many other navies subsequently incorporated steam vessels into their fleets, initially only on

MISSISSIPPI MAN

American author Mark Twain (1835–1910) used his own pre-Civil War childhood on the Mississippi as the basis of his two most famous novels, *The Adventures of Tom Sawyer* (1876) and *The Adventures of Huckleberry Finn* (1884). In Twain's day, hundreds of paddlewheel steamers plied the river. The writer had many different jobs during his lifetime, including typesetter, journalist and gold prospector, but it was his spell as a pilot on one of these riverboats that gave him his pen name: 'Mark Twain' was boatmen's slang for a depth of two fathoms measured by sounding line. His real name was Samuel Langhorne Clemens.

auxiliary vessels. The first Royal Navy steamships were the tugs *Comet* and *Monkey*, which entered service in 1822.

Even so, the potential of the first generation of steamships was hampered by paddlewheels, which were inefficient in anything but the calmest of seas. The advent of two key technologies ushered in a second phase of development that eventually saw steam supplant sail both in frontline naval service and in merchant shipping: screw propellers and iron hulls. The first vessel to incorporate both was Isambard Kingdom Brunel's revolutionary SS *Great Britain*, which was launched in 1843. Over the following decades, sail gave way to steam in virtually every kind of vessel afloat, from trawlers to passenger liners to battleships, and the gross tonnage, size and speed of ships increased exponentially.

In the footsteps of Huckleberry Finn
The Natchez *(above), built in 1975, is a replica of a traditional sternwheel paddle steamer. She takes tourists on cruises up and down the Mississippi from New Orleans.*

Heyday of the steamship
In the 1920s and 1930s, steamship lines vied with one another to win passengers for their services to destinations near and far. This 1925 poster advertises the Baltic services operated by the French Worms Line.

STEAMBOAT BILL AND WILLIE

Two very different American movies of 1928 featured Mississippi steamboats. The full-length silent feature film *Steamboat Bill, Jr*, released in May 1928, starred Buster Keaton as a hapless young college graduate trying to make his way as a riverboat captain. Six months later, in November, the Walt Disney cartoon character Mickey Mouse made his first appearance in the premiere of *Steamboat Willie* at New York's Colony Theater. The year 1928 marked the transition from silent movies to talkies, and Disney's animated film was notable for being the first cartoon to have synchronised sound.

Lifeboats c1789

During the 18th century, some 5,000 ships a year called at the port of South Shields at the mouth of the River Tyne to load up coal from the mines around Newcastle. Shipwrecks were commonplace. In 1789 tragedy struck the coaler *Adventure*, which foundered on the Herd Sands just 200 metres offshore. A crowd of onlookers, powerless to help, watched as her captain – a local man – and his seven shipmates leapt into the rough seas and drowned. In response, a group of philanthropic local worthies organised a public competition, with a prize of 12 guineas (£12 12s), to anyone who could come up with a practicable design for a lifeboat 'capable of negotiating heavy surf in extremely shallow waters'.

The winning entry was built by Henry Greathead, a local shipwright. Greathead is generally credited as being the father of the lifeboat, though a number of other inventors had already constructed unsinkable or self-righting boats: in 1775, a French official, Monsieur de Bernières, launched such a craft on the Seine. Lionel Lukin, a London coachbuilder, built a boat with watertight compartments in 1783, while in 1789 William Wouldhave submitted a design for a lifeboat made of copper to the competition won by Greathead.

Greathead's lifeboat, the *Original*, was shaped to cut through the waves and was made more buoyant by having cork inside and out. She soon proved her worth, saving 200 people in 1797 alone. Within 20 years, Greathead lifeboats were in use throughout the British Isles, as well as in Denmark, Sweden and Russia. But they were heavy and difficult to launch and required a crew of 24 oarsmen. From 1850 onwards, they were replaced with new, lighter designs incorporating both sails and oars. Motorised lifeboats only became widespread after the Second World War.

A TRAGIC GAMBLE

When the RMS Titanic hit an iceberg and sank on her maiden voyage to New York, she did not have sufficient lifeboats on board for all the passengers and crew. In the event, only 711 of her total complement of 2,208 were saved. She carried enough boats to satisfy the legal requirements of the day, but these had been overtaken by the rapid growth in the tonnage of liners and the number of passengers they could accommodate.

Unbelievable loss *The 'unsinkable'* Titanic *sank on 14 April, 1912 (below). Subsequently, a British Board of Trade enquiry made it mandatory for every ship to carry sufficient lifeboat accommodation for everyone on board.*

Braving the waves *Seen here (left) in an engraving from the* Encyclopaedia Londoniensis *of 1810–29, Henry Greathead's lifeboat,* Original, *was in service for 40 years. She was finally wrecked on rocks, with the loss of two of her crew.*

False teeth 1788

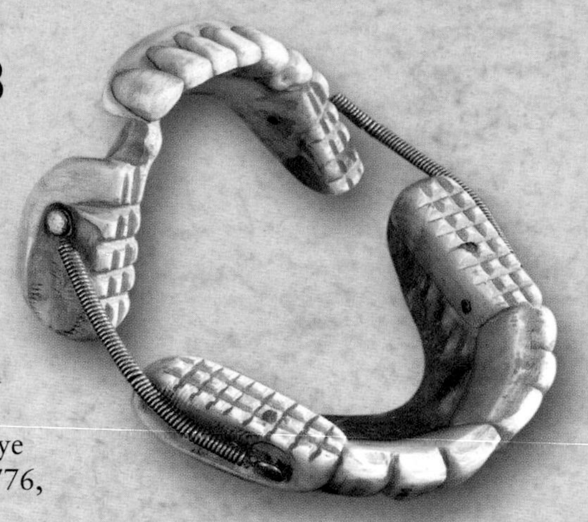

Artificial mouthful
A set of false teeth made of ivory and gold (right). They would not have been recommended by Parisian dentist Dubois de Chémant. He published a paper on the subject in 1788, which later appeared in English as A Dissertation on Artificial Teeth: Evincing the advantages of Teeth made of Mineral Paste, over every denomination of animal substance.

Crude false teeth first appeared in the 7th century BC: made from ivory or bone, they were fragile and unhygienic. The earliest known full set of false teeth – carved from bone and attached with gut to a hinged side piece – were dug up in a Swiss field and date to the 15th century. False teeth as we know them today were the invention of two Frenchmen: Alexis Duchâteau, an apothecary from Saint-Germain-de-Laye who produced a prototype set of porcelain teeth in 1776, and a Parisian dentist, Nicolas Dubois de Chemánt.

Camembert cheese 1791

Mass production
Engraving from Louis Figuier's Marvels of Industry *(left), published 1873–6, showing a worker in a Normandy creamery ladling curds into moulds for ripening into Camembert. From around 1890, the characteristic wooden box made it a great export hit.*

Legend has it that Camembert cheese was first made in 1791, when a priest named Charles-Jean Bonvoust took refuge in a Normandy farmhouse owned by one Marie Harel while fleeing the French Revolution. In return for her kindness, he gave her the recipe for the soft cheese that had already made Brie, his native region, famous. She used the recipe to improve a local cheese that had been around since the 1600s. A further improvement came *c*1910, when the addition of the ripening mould *Penicillium candida* turned the rind's original blue-grey colour to the chalky white familiar today. Since then, Camembert has been fêted as one of the world's great cheeses.

REGIONAL DIVERSITY

Both Britain and France pride themselves on the variety of cheeses produced in different regions, many of which have 'Protected Designation of Origin' status under EU law. Yet the great French statesman Charles de Gaulle once exclaimed: 'How can anyone govern a nation that has 246 different cheeses?'

AN INTERNATIONAL CHEESEBOARD

Some of the world's great cheeses have been enjoyed for many centuries. Cheddar, for example, was first recorded in the late 12th century. Traditionally made in the Cheddar Gorge region of Somerset, its great popularity led to it being imitated worldwide. The hard Italian cow's milk cheese Provolone has also travelled far. Stilton, originally from a farm near Melton Mowbray in Leicestershire, dates from the late 17th century. Swiss Emmental can trace its origins back to the mid-15th century, while the salty French blue cheese Roquefort has been matured in caves in the village of the same name in southern France since at least the early 15th century.

A fine spread
Cheeses in a still-life painted in 1625 by Flemish artist Roelof Koets the Elder.

The guillotine 1792

On 10 October, 1789, Dr Joseph Ignace Guillotin recommended to the French Revolutionary Constituent Assembly that 'all cases of capital punishment … should be of the same kind – that is, decapitation – executed by means of a machine'. Guillotin's main concern was to alleviate the undue suffering that condemned prisoners had been subjected to in pre-Revolutionary France. The 'humanitarian' execution device that now bears his name was actually conceived by the surgeon Antoine Louis, who in turn got the idea from a similar apparatus called the 'maiden', used in Scotland until the late 17th century: 'Two upright posts surmounted by a crossbeam, from which a trigger mechanism releases a blade that falls onto the neck of the condemned man.'

LAST EXECUTION

In the 19th and early 20th centuries, the guillotine was the main means of execution in several European countries, including Greece and Germany. The last person to be guillotined in France was the Tunisian murderer Hamida Djandoubi, in 1977. The guillotine remained France's legal method of execution until 1981, when President François Mitterrand led a successful campaign to abolish the death penalty.

After being tested on live sheep, bales of straw and human cadavers, the guillotine was first used in earnest on 25 April, 1792. On the Place de Grève, executioner Charles-Henri Sanson dropped the razor-sharp steel blade onto the neck of highwayman Nicolas-Jacques Pelletier. Parisians found the spectacle too clinical and assembled at the same spot the next day to chant 'Bring back the gallows!'

The guillotine became the quintessential symbol of the Reign of Terror that followed the French Revolution. Appalled at the bloodshed, Guillotin withdrew from public life and died in obscurity in 1814. The novelist Victor Hugo later said of him: 'Some men are just plain unlucky. Christopher Columbus didn't manage to get his name attached to his discovery, whereas Guillotin couldn't disassociate his from his invention.'

Madame Guillotine
A 19th-century model of a guillotine (left). According to its eponymous inventor: 'The blade falls, the head is severed in the blink of an eye, and the victim dies instantly. The most he feels is a swift breath of cool air on the nape of his neck'.

GRIM HUMOUR

The guillotine won a variety of nicknames, such as the Louison or Louisette (after its original inventor), the 'national razor', the 'widow', the 'tie of the Capets' (the family name of the last French royal family) and the 'mill of silence'.

Public spectacle
The last public execution in France took place on 17 June, 1939, when serial killer Eugen Weidmann was guillotined outside the prison at Versailles.

THE METRIC SYSTEM – 1792

Getting the measure of things

Revolutionary France, which prided itself on promoting equality and uniformity, was quick to implement a new system of weights and measures. The long-running story of metrication takes many unexpected twists and turns, but it began with an attempt by two French astronomers, Delambre and Méchain, to establish the metre as the standard unit of linear measurement.

Tailor-made for trade
Before widespread adoption of the metric system, the plethora of different measurement systems hampered both international trade and scientific research. This 15th-century Italian fresco shows tailors at work.

On 10 August, 1792, the astronomer Jean-Baptiste Delambre stood in the bell tower of the church at Dammartin, 35 kilometres outside Paris, waiting patiently for his assistant Michel Lefrançais to light a brazier on Montmartre hill. But the pre-arranged signal – which he was planning to use to take a sighting with his theodolite – never came. That very day, tens of thousands of Parisians attacked and pillaged the Tuileries Palace and set the capital ablaze. Lefrançais thought it wise not to attract the attention of the angry mob.

This was just the first of many setbacks for Delambre and his fellow astronomer, Pierre Méchain. It would take seven years for them to complete the task they had been set – to measure the length of the meridian from Dunkirk to Barcelona. The ultimate aim of the exercise was to define a new unit of measurement as part of root-and-branch reform of weights and measures in France.

CONFUSION REIGNS

Before the spread of the metric system, a plethora of different national and regional units of measurement existed for working out length, area, volume and weight. The ell, for example, was used for measuring bolts of woven cloth. It was based on the approximate length of a man's arm, but it differed widely between countries: the English ell was 114cm, the Scottish ell 94cm, the Polish ell 79cm, and the Flemish ell just 69cm. Likewise the perch – from the *pertica*, the standard Roman unit for measuring tracts of land, equivalent to 3m – varied from place to place and over time. In England, where it was also called the 'rod', it measured 18, 20, 22 or 24 feet (5.5 to 7.3m). Similarly, in medieval France the perch was 3m in Lorraine, 5.4m in Paris and 7.3m in Brittany. For defining areas of arable land, it was not uncommon for farmers to use seed measures rather than geometric units – that is, the quantity of grain that was required to sow a field.

In pre-Revolutionary France, the situation was particularly confusing. British agriculturalist Arthur Young, who travelled there in 1787–9, reported that: 'In France, the infinite complexity of the measures exceeds all comprehension. They differ not only in every province, but in every district and almost in every town.'

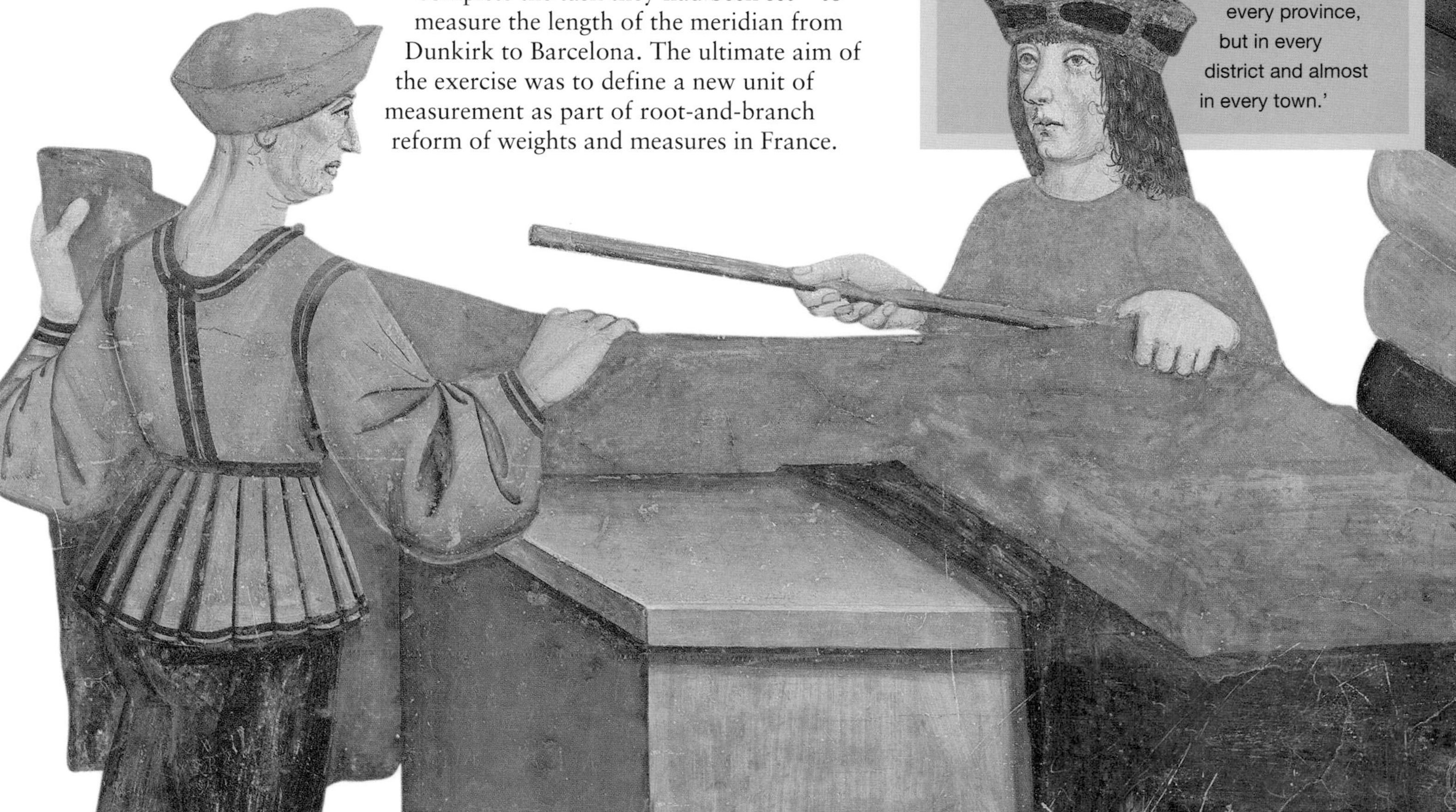

Triangulating the arc
An engraving of 1873 (above) depicts Méchain and Delambre measuring a section of the meridian arc.

Rational and universal

Official dossiers from 1789 recording public grievances are littered with calls for an end to the confusion surrounding units of measurement in France. Even so, those charged with reforming the system were met with hostility from the very people they were trying to help. As surveyors criss-crossed the country taking sightings, it was not uncommon for them to find themselves suspected of counter-revolutionary espionage or even of being snooping tax inspectors.

The move to standardisation of weights and measures was driven by political as well as economic ulterior motives. Leading advocates of reform saw it as a key part of a wider agenda to promote social equality and financial probity. In the drive to create a centralised state funded by a system of equitable taxation levied on business and property, rationalisation of weights and measures went to the heart of French national unity. The mathematician Auguste-Savinien Leblond coined the term 'metre' (from the Greek *metron*, meaning 'measure') for the new unit of length, and to make it immutable and universal it was decided to base the unit on the Earth's dimensions. The academicians were then determined to persuade their countrymen to adopt their new unit of linear measurement.

Man and measure
A portrait of Jean-Baptiste Delambre (1749–1822) by Henri Coroenne (above).

Diligent surveyor
Pierre Méchain (1744–1804, inset) was responsible for measuring the southern part of the Dunkirk–Barcelona meridian. Doubts about the accuracy of his measurements led him to return to Spain in 1804, where he died of yellow fever.

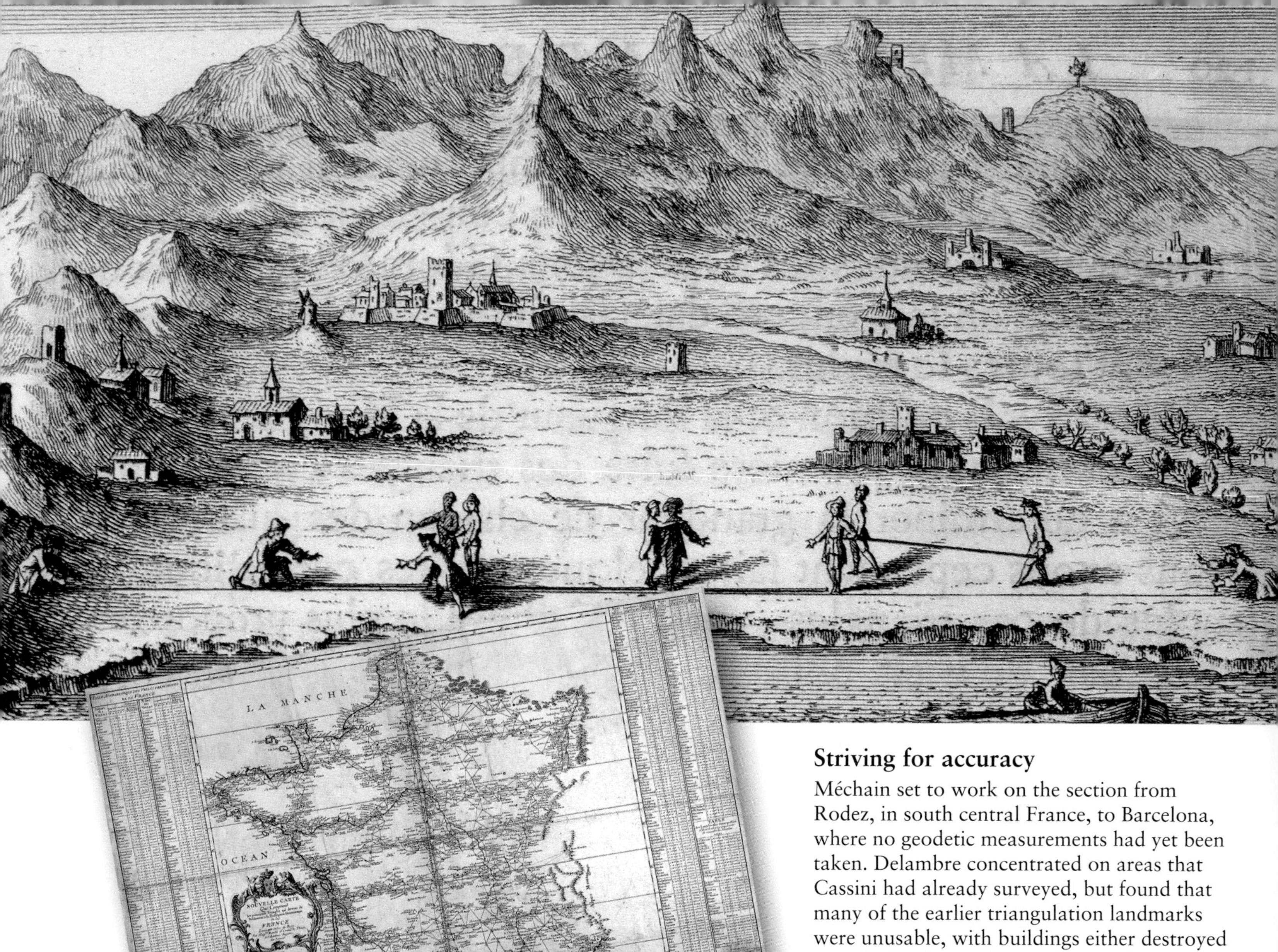

Lie of the land
A plate from César Cassini's 1744 publication The Meridian of the Royal Observatory of Paris, *showing geodetic surveyors at work (top). The work summarised the conclusions of a survey of the meridian through Paris undertaken in 1739–40. The map of France (above) published by Cassini and Maraldi in 1744 was the first accurate cartographic portrayal of the country.*

Accordingly, in March 1791, the National Assembly settled on the meridian as the basis of the future system. The plan was to measure a quarter meridian, or quadrant – the arc from the Pole to the Equator – then divide this by 10 million. The base unit that would emerge from this work was both easy to use and, as luck would have it, close to the Parisian ell, a unit already familiar to most French people.

The physicist Jean-Charles Borda suggested that the measurement be made using the quadrant that passed through Paris. This was a logical choice, since it was the only arc on the Earth's surface that had already been partially surveyed. In 1740 César François Cassini, the latest scion in a family of astronomers who had run the Paris Observatory since its inception in 1671, measured the distance from Dunkirk to Perpignan. Completion of the task was entrusted to Delambre and Méchain.

Striving for accuracy

Méchain set to work on the section from Rodez, in south central France, to Barcelona, where no geodetic measurements had yet been taken. Delambre concentrated on areas that Cassini had already surveyed, but found that many of the earlier triangulation landmarks were unusable, with buildings either destroyed or so badly damaged as to be unsafe. Also, tree growth in the intervening years and new buildings often obscured the sight line from one trig point to the next.

The method of triangulation used to measure distance was simple – and so effective that it has only latterly been supplanted by geosurvey satellites. First described in 1445 by the Italian Renaissance painter and architect Leon Battista Alberti, the technique entailed establishing a network of imaginary triangles along the entire length of the distance to be surveyed. The vertices of these triangles were provided by vantage points along the route, such as church bell towers or hills, each visible from the preceding one. Each triangle had one side in common with its neighbour, and their bases were aligned with the meridian. The completed network formed a chain straddling the meridian.

From the vantage points, or 'stations', surveyors would take sightings to determine the angles of the triangles. Then, by measuring the length of one side of each triangle (the base) – an operation achieved by the simple method of laying straight rules end to end along the ground – the length of the remaining two sides could be calculated. From this, the scientists worked out the length of the meridian in the section being surveyed.

In Cassini's day, instruments measured angles only to the nearest 15 seconds. Since then, Borda and the instrument maker Étienne Lenoir had created a geodetic survey instrument – the 'repeating circle' (1784) – that was accurate to the nearest second. Méchain took accuracy to the point of obsession. Like Delambre, his starting point for ascertaining the length of the arc of the meridian – and hence the total distance covered by the quadrant – was the precise latitude of the furthest point of the section that he was due to survey. This was the castle of Mont-Jouy, overlooking Barcelona. During the winter of 1792–3, Méchain made no fewer than 1,050 sightings of six stars from there, which were his reference points for determining latitude. He repeated each sighting 10 times. The findings all tallied, with one exception: a tiny discrepancy of 4 seconds of arc for the star Zeta in the constellation Ursa Major. This represented a variation of just 0.01 per cent in the total length of the meridian arc between Dunkirk and Barcelona.

Méchain, who had discovered 11 comets and was renowned for the reliability of his planetary tables, refused to accept this error, which he ascribed to the effects of light refraction. But the outbreak of war between France and Spain in May 1793 prevented him from returning to Mont-Jouy. The following winter, he took 10,000 or so sightings from the terrace of an inn where he had been billeted by the Spanish authorities. These only served to alarm him more, as the latitude measurements differed by 3.2 seconds of arc from those originally obtained at nearby Mont-Jouy. Tormented by the discrepancy, Méchain gave up surveying for three years.

Elusive perfection

Meanwhile, Delambre was beset by political interference. In January 1794, the Committee of Public Safety summarily cancelled the meridian project. Republican France's new strong men wanted to institute the metric system without more ado. Delambre and Méchain's work was taking too long, so the Committee voted to accept the provisional metre measurement that the eminent mathematicians Borda, Pierre Simon de Laplace and Joseph Lagrange had worked out using Cassini's

FROM THE METRE TO THE GRAM

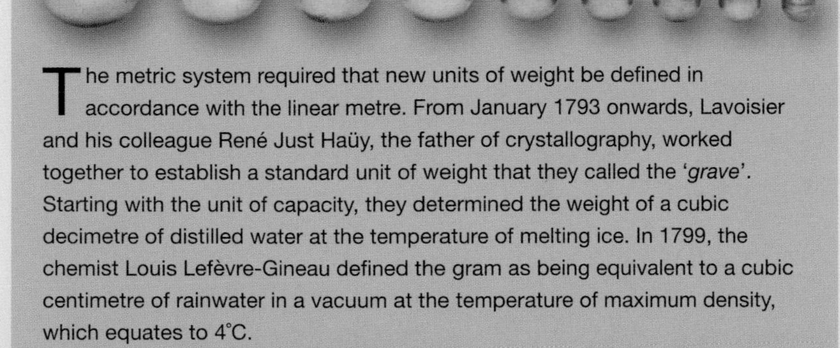

The metric system required that new units of weight be defined in accordance with the linear metre. From January 1793 onwards, Lavoisier and his colleague René Just Haüy, the father of crystallography, worked together to establish a standard unit of weight that they called the '*grave*'. Starting with the unit of capacity, they determined the weight of a cubic decimetre of distilled water at the temperature of melting ice. In 1799, the chemist Louis Lefèvre-Gineau defined the gram as being equivalent to a cubic centimetre of rainwater in a vacuum at the temperature of maximum density, which equates to 4°C.

Another system *A set of troy weights (above), made by London instrument maker Robert Bate in 1824. The Troy ounce is a unit of mass, dating from medieval times, that is still used in England to weigh precious metals: 32 Troy ounces make 1 kilogram.*

Repeating circle
This geodetic survey instrument used two telescopes, mounted on a shared axis, that rotated independently around a graduated circle. By sighting the telescopes on separate fixed points, locking them to take a reading, then repeatedly resighting them, the surveyor could obtain a highly accurate final measurement of angle.

Leading the way
A postcard of 1880 (above) proclaims France as the pioneer of metrication.

1740 figures. The decimal basis of the metric system was retained, as was the now-familiar nomenclature for the units that derive from the metre, the centimetre and millimetre.

In June 1795, with the Reign of Terror over, Delambre and his assistants were able to resume work and complete the remaining triangulations. Three years later, in November 1798, scholars from the Netherlands, Denmark, Switzerland, Italy and Spain gathered in Paris for the first international science congress. Their task was to establish the standard measurement of length on the basis of Delambre and Méchain's calculations.

When they scrutinised the results, the International Metric Commission was astonished to find that the curvature of the Earth's surface was not only more elliptical, or eccentric, than previously thought, but also that it varied from segment to segment. Contrary to the hypothesis on which the choice of the Paris meridian was based, all meridians were not equal in length. Believing that César Cassini had calculated the degree of eccentricity more accurately, the Commission ignored all of Delambre and Méchain's work and reverted to the 1740 data to extrapolate the arc of the Dunkirk to Mont-Jouy meridian to the quadrant. It is now known that the metre standard proclaimed in June 1799 – cast in platinum, dedicated to 'all times and all men' and deposited in the archives of the French Republic – is short by 0.2 millimetres.

A fruitful error

The discrepancy was not the result of an error by Méchain but due instead to the imperfect shape of the globe. Méchain had presented his findings without revealing his later series of conflicting measurements, which were not taken into account in the final calculations. In other words, he had modified his findings to make them appear more precise and consistent. Delambre did not discover this until after his colleague's death. Thereafter, Delambre spent many years piecing together the original data from Méchain's archives and finally published them in a treatise entitled *The basis of the decimal metric system, or the measurement of the arc of the meridian.*

The third and final volume of this work appeared in 1810, two years before Napoleon abolished the metric system in February 1812 (it would be reinstated in 1837). Delambre presented Méchain's difference of 3.2 seconds of arc in his measurements not as an error but as a discovery, and this opened up fruitful new lines of enquiry. In 1828, mathematician Jean Nicolas Nicollet introduced the concept of

ALMOST UNIVERSAL

Over time, almost all countries adopted the metric system, with two notable exceptions. In Britain, a Parliamentary committee recommended metrication as early as 1862, but it was not until 1965 that the policy was adopted and it was 2000 before shopkeepers were legally required to sell goods by the metre or kilo – a measure since relaxed. Speedometers and road signs are still in miles, as they are in the USA despite the Metric Conversion Act of 1975. At 1,609m the terrestrial mile differs from the nautical mile (1,852m) and this in turn is slightly longer in Commonwealth countries (1,853.18m) than in the UK. Switching between systems can cause problems: the loss of NASA's Mars Climate Orbiter in 1999 was due to subcontractors working in different units. The error put the craft 96km below its intended orbit, causing its destruction in the planet's turbulent atmosphere.

Ready reckoner
The Maurand disc calculator of 1863 contained 35 double-sided conversion discs.

systematic error in scientific analysis, which showed that Méchain's discrepancy was not really a discrepancy at all. Henceforth, scientists began to factor in the uncertainty principle, and quantify it through statistical analysis, a discipline then in its infancy.

At long last, the International Metric Commission had all the tools it needed to implement the new system. In 1872, 30 scientists from Europe and America met in France to oversee the introduction of new standard measurements. The metre in the archives was replaced by a new prototype made of a harder, platinum–iridium alloy, and an International Bureau of Weights and Measures was established at Sèvres in 1875.

Measure of success
British athlete Denise Lewis on her way to gold in the heptathlon at the summer Olympic Games in Sydney in 2000. The metric system was adopted for recording Olympic sporting achievements as early as 1908.

SHIFTING DEFINITIONS

Originally representing ten-millionths of the distance between the North Pole and the Equator, the metre was redefined in 1960 in terms of the wavelength of light emitted in a vacuum by the isotope krypton-86. The International System of Units (SI units) sets the speed of light in a vacuum, as measured by an atomic clock, at 299,792,458 metres per second. It is an irony of history that the early French advocates of the metric system considered using the length of a pendulum beating one second as the base of the metre, but rejected it because it relied on a unit of time. But the time principle did at least have the merit of winning sceptical Anglo-Saxon cultures round to the idea of the metric system.

GAS LIGHTING – 1792

Illuminating the cities

Up to the end of the 18th century, people lit their homes, as they had for centuries, with oil lamps and candles. Street lighting in cities and towns was non-existent. Brave pioneering work by inventors led to the introduction of coal gas, which became the principal source of lighting until the advent of electricity.

A violent end
Philippe Lebon, pioneer of gas street lighting in France, was assassinated on 2 December, 1804, while crossing the Champs-Élysées in Paris to attend Napoleon's coronation as emperor.

The urge to lengthen the day with artificial lighting was not new: oil lamps were used in many ancient and even prehistoric societies, and candles were in use from the Middle Ages. Lamps were continually improved throughout the 18th century, but they had one major flaw: the fuel they burned was mostly rapeseed oil, which gave off a strong and all-pervading stench.

A brilliant idea

In 1779, Scottish engineer William Murdock was sent to Redruth in Cornwall by his employers, the steam engine entrepreneurs James Watt and Matthew Boulton. His job was to supervise the installation of steam engines to work pumping equipment in the tin mines. While there he began experimenting with harnessing the gas given off by burning coal. He set up an iron retort in the backyard of his home in Redruth and conveyed the gas via a pipe into the cottage. On 29 July, 1792, he succeeded in lighting a flame inside the room.

Meanwhile, a French engineer and chemist Philippe Lebon had put forward a proposal to produce a combustible gas for lighting on a large scale. His gas distillation process, patented in 1799, involving heating wood in a retort. As the wood smouldered, it released a flammable, foul-smelling gas that he called 'hydrogen', though in fact it was a complex mixture of different gases. He duly installed his system in the Hotel Seignelay in Paris, the first recorded instance of a public building lit by gas, but he failed to attract investors. Three years later, he was stabbed to death in mysterious circumstances.

Murdock's coal-gas was as foul-smelling as Lebon's wood-gas, due to the presence of sulphur, but around the turn of the 18th century he succeeded where Lebon had failed. He installed his lighting system at the Soho Foundry in Birmingham and a textile mill in Manchester soon followed suit.

Professional lamplighter
From the early 19th century, lamplighters were a familiar sight in city streets, lighting and extinguishing the gas street lamps.

COAL TAR AND COKE

The first person to demonstrate that gas could be obtained from coal was John Clayton of Wigan, Lancashire, in around 1684. The process was industrialised around a century later to produce coal tar, which was used in caulking the hulls of wooden ships and boats to make them watertight. The gas that was given off was long viewed as an unwanted by-product, which was flared off. The solid coke residue – the coal equivalent of charcoal from wood – was later put to good use in the iron smelting industry.

The growth of street lighting

In 1804 the British inventor Frederick Albert Winsor (born in Germany as Friedrich Albrecht Winzer) patented a furnace for the production of coal gas, and went on to found the National Light and Heat Company. In 1812 the company was granted exclusive rights by the Crown to produce coal gas for public lighting. Over the next 25 years it set about installing street lamps throughout London. Despite public concern at the danger of explosion, the world's first gasworks was built in Westminster in 1812. Three years later, Winsor set up a

Safer streets
The first town to be equipped with lampposts burning coal gas was Freiberg in Saxony, Germany, in 1811. Other cities soon followed, including London (1812), Baltimore (1816) and Paris (1818). Gas lamps feature in many paintings of the period. This one of 1900 by the Swedish artist Aron Gerle is simply entitled A Street at Night.

gas-lighting network in Paris, which came on stream in 1818. Gasometers – a term for gas storage tanks coined by Murdock – and street lamplighters soon became familiar sights in European cities. The popularity of the lighting was due in large part to British engineer Samuel Clegg, who invented a purification process that cleansed 'town gas' of its foul odour.

A brighter light

The incandescent gas mantle, which generated a much brighter light when heated by gas than a simple gas flame, was invented in 1881 and named the 'Clamond basket' after its French inventor. Devised for use in the home, it was made from a mixture of magnesium hydrate, magnesium acetate and water squeezed through holes in a plate to form thread. The entire delicate 'structure' was supported by a platinum wire cage. Later versions used cotton thread soaked in a variety of oxides and were the invention of Carl von Welsbach, a former student of Robert Bunsen.

Fuel saving
Even before the German occupation, the French began experimenting with hydrogen gas-powered cars. Here, a Parisian taxi driver fills his vehicle's gas-generator with anthracite coal in 1938.

RUNNING CARS ON GAS

The gas generation process enjoyed an unexpected revival during the German occupation of France in the Second World War, from 1940 to 1944. With petrol severely rationed, cars and lorries were fitted with gas-generators fuelled by wood, coal or charcoal. Burners heated a closed chamber containing the fuel, generating hydrogen to run the vehicle's engine. A similar ingenious solution was tried in Britain at the same time; digesters using human waste or animal manure produced methane, which was stored in huge gas bags on the roofs of vehicles.

THE OPTICAL TELEGRAPH – 1793

The dawn of telecommunications

In an effort to improve the range and speed with which information could be transmitted, the French engineer Claude Chappe devised the optical telegraph, an ingenious network of aerial relay stations. Chappe's invention represented the first practical telecommunications system.

A signal innovation *Claude Chappe's notebook of c1794 contains extensive descriptions and sketches of his semaphore signalling system. Shown here (right) are drawings of four different signal positions.*

In around 1790, a letter sent by stagecoach took four days to travel the 500 kilometres (300 miles) between Paris and Strasbourg. Ten years later, a message covered the same distance in two hours; by 1850 the time was under six minutes. What brought about this radical change was a groundbreaking invention by Claude Chappe (1763–1805). Chappe started off training for the priesthood, but soon found himself far more interested in physics and engineering, and in particular the question of how technology could be applied to transmitting messages over long distances.

Signalling with semaphore

The communications system that he envisaged was based on a network of semaphore stations. The signalling mechanism on these comprised a large transverse bar (louvred to reduce wind resistance) mounted centrally on a pylon. This cross-arm – known technically as the 'regulator' – could be set in four positions: vertical, horizontal, and tilted either left or right. At each end of the arm, which was 4.65 metres long, was pivoted a smaller, 2-metre-long arm: the 'indicator'. The indicators could be set at an acute, obtuse or right angle.

A complex system of levers, axles, counterweights, ropes and pulleys enabled the operator to move all three parts of the mechanism independently of one another. The various positions of the regulator and its two indicators corresponded to different numbers, which in turn, via a code, denoted a particular letter, word or phrase. Messages were read through a telescope from the next relay station; these were set several miles apart, on natural vantage points, such as hills or mountain ridges, or alternatively on top of church towers or other tall buildings. The main drawback was that the system was inoperable at night – lights were tried but found to be impractical – and also when it was foggy, raining heavily or snowing.

At the mercy of the mob

Chappe's first public demonstration of his semaphore chain took place on 2 March, 1791, between his home town of Brûlon and

AENEAS' HYDRAULIC TELEGRAPH

Since ancient times, drums, trumpets, the human voice, beacons and smoke signals have been used to send messages over long distances. In the 4th century BC Aeneas Tacticus, a Greek writer on military matters, combined fire and water in the first optical communication system. His hydraulic telegraph comprised identical large terracotta jars sited on the tops of hills; each jar was filled with water, and a vertical graduated rod floated inside. The rods were inscribed with various predetermined codes indicating specific events, such as the arrival of ships, infantry or cavalry. To transmit a message, the sending operator held up a torch as a signal to the receiver, who responded in kind; once the two were synchronised, they put out their torches and simultaneously pulled out bungs plugging standard-sized holes drilled in the bottom of their jars. The jars emptied until the water level reached the desired code, at which point the sender relit his torch, prompting the receiver to plug his jar and read off the relevant information.

Nationwide network
A lithograph from the 19th-century magazine Le Petit Journal *(left) shows Chappe demonstrating his optical telegraph. By 1844, the telegraph network had expanded to 556 stations connecting 29 French cities over a total distance of 5,000 kilometres.*

Parcé, 15 kilometres away. It took nine minutes to transmit 26 words. The Parisian authorities subsequently asked him to install one of his devices on the Étoile gate on the city's defensive wall, but no sooner had he erected it than the mechanism was pilfered by timber thieves.

With the support of his brother Ignace, a member of the Legislative Assembly, Chappe presented his invention to the National Convention in March 1792. The revolutionary government was impressed: they saw the device as a possible way of 'facilitating communications with a besieged town', and so earmarked funds for testing the system over a 'suitably lengthy chain of transmission'. Accordingly, Chappe built a signalling station at Ménilmontant, but it was razed to the ground by the working-class radicals known as *sans-culottes*, who suspected that it was going to be used to communicate with the imprisoned French royal family.

THE FIRST INSIDER TRADERS

The optical telegraph played a key role in an infamous 19th-century swindle. In 1835, twin brothers François and Louis Blanc, speculators on the Bordeaux stock market, bribed a telegraph operator in Tours to make pre-arranged errors in transmissions to signal dramatic rises or falls in the price of government securities on the Paris stock exchange. They used this information to make profits in Bordeaux before rival investors got the news from Paris by stagecoach post five days later. And because there was no law expressly forbidding the use of the telegraph for private messages, the Blancs escaped conviction.

Many permutations
A model of a Chappe telegraph of 1873 (below), with a table of signal positions and their corresponding letters and numbers. Each arm had seven positions and the cross-arm four more, creating a code of 196 combinations.

NAME CHANGE

Chappe originally named his invention the 'tachygraph', from the ancient Greek words for 'speed' and 'write', but he later changed the name to 'telegraph' from Greek *tele-*, meaning 'far', and *graphein*, 'to write'.

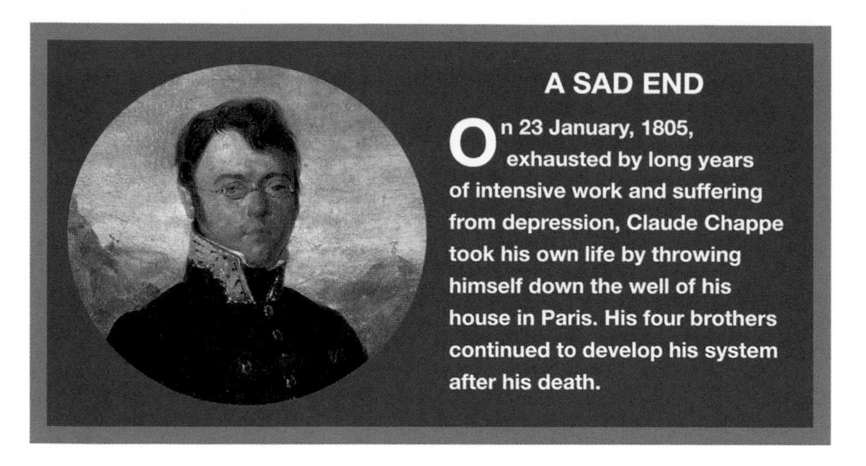

A SAD END

On 23 January, 1805, exhausted by long years of intensive work and suffering from depression, Claude Chappe took his own life by throwing himself down the well of his house in Paris. His four brothers continued to develop his system after his death.

Military application

The Assembly recognised the potential of the system at a time when France was under threat from foreign powers, and so, undeterred by the setback, they ordered Chappe to construct a trial chain of stations from Ménilmontant to Saint-Martin-du-Tertre. It was a success and the authorities conferred on him the title 'telegraph engineer' in July 1793. Chappe's next commission was to build a relay line from Paris to Lille, with 16 stations covering the 192 kilometres (120 miles) from the war front. On 1 September, 1794, the war minister Lazare Carnot proudly announced to the Convention: 'The Austrians are in full retreat. The fortress at Condé-sur-Escaut, which they held until recently, has just fallen to our forces. We were informed of this development within the hour by the Chappe telegraph ...'

Success elsewhere

Around the same time as Chappe's telegraph, the Swedish inventor Abraham Edelcrantz and Lord George Murray in Britain were also experimenting in optical telegraphs, with systems involving shuttered wooden panels rather than cross-arms. A British version of the telegraph, with six wooden shutters capable of 64 different combinations, was ordered by the Admiralty in 1795. The first station, at a house in West Square, Kennington, was on a route linking the Admiralty with Chatham and Deal in Kent. Other routes ran to Yarmouth, Portsmouth and Plymouth. One station has been preserved and can still be seen at Chatley Heath in Surrey.

Other countries soon followed suit, including Prussia, the United States, China and Tsarist Russia. By the late 1830s, though, a new communication system was beginning to emerge: the American inventor Samuel Morse came up with the electric telegraph. With the spread of this more effective technology, optical telegraph systems faded into obscurity from 1850 onwards, after more than half a century of service.

Old but effective
During the Crimean War (1853–6), communication between units of the allied armies from France, Britain and Turkey were conducted with the new electric telegraph. Yet Chappe's tried-and-tested semaphore system was also used, as shown in this 1888 engraving of a French camp in the Crimea.

The cotton gin 1793

Although the American Eli Whitney Jr trained as a lawyer, his true passion was for engineering and he was responsible for an invention that revolutionised cotton spinning. In 1793, while staying at a large plantation in Georgia called Mulberry Grove, he learned about the problems involved in processing cotton. In particular, once the workers had picked the raw bolls, they had to separate the white fibre (the part that is made into yarn and cloth) from the seeds. This labour-intensive and tedious task was done by hand, and it seriously slowed down production.

Some months later, Whitney produced an ingenious machine that comprised a hand-cranked, rotating horizontal cylinder, studded with spikes that passed through the narrow bars of a mesh grid. W en the cotton was fed through the device, it was effectively 'combed' – the spikes caught the fibres, while the seeds were too big to pass through the mesh and dropped into a hopper below.

From cotton to guns

Whitney's cotton gin helped American cotton production to leap from 10,000 bales in 1794 to 4 million in 1861 (a bale weighs around 230 kilos). It made his name as an inventor but, always more an engineer than a businessman, he only patented the gin in March 1794, by which time others had copied the design. He went on to make his fortune mass-producing muskets with interchangeable parts, a system of manufacture that pre-empted Henry Ford's assembly line by a century.

Whitney's wonder
The cotton gin ('gin' is short for 'engine') underwent few modifications over the many decades it was in use, but one improvement was in the power supply. It developed from a hand-cranked machine, like the one shown here (left), to being driven first by horse-power and then by water mills.

ESCALATING DEMAND FOR SLAVES

Whitney's cotton gin may have done away with the drudgery of seeding cotton by hand, but it did little to alleviate slavery. In fact, its effect was quite the contrary. The cotton plantations in the southern states of the USA depended on the institution of slavery. As production boomed, the numbers of slaves increased – from 500,000 in 1750 to 1.5 million in 1820. Slavery was only abolished in the United States in 1865, at the end of the American Civil War.

The corkscrew 1795

The increasing use, from the 17th century onwards, of corks as bottle-stoppers gave rise to the corkscrew. In 1795, Oxford clergyman Samuel Henshall had the idea of inserting a disc between the screw (the 'worm') and the shaft ('shank') of the corkscrew. This simple mechanism, which became known as the 'Henshall button', stopped the worm from boring too deeply into the cork and, once it reached the bottle neck, broke the tension between the cork and the glass, so forcing the cork to move up with the twisting of the crosspiece handle.

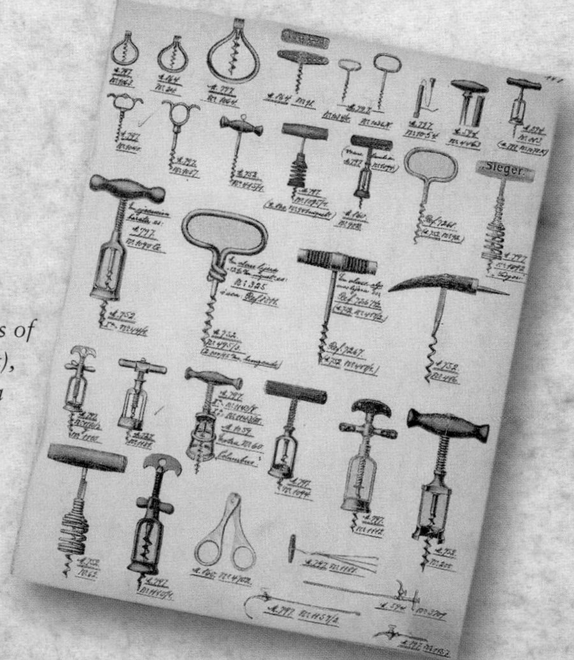

Many styles
Different models of corkscrew (right), as presented in a domestic goods catalogue of c1890–1910.

Keeping food fresh

Nowadays, canned food is so commonplace it is almost scorned, but at its introduction this way of preserving food was hailed as a boon to humanity. The method of preservation was invented in 1795 by a French confectioner using bottles, transferring to tin a few years later.

Bringing up baby
The sterilised nature of canned produce made it ideal for baby food. The child on this 1920s label from the Chevallier-Appert cannery (right) eagerly announces 'I Want Some Too!'

Earlier methods of preserving food long-term, some of which date back into prehistory, include drying, salting and smoking. Other methods known in the 18th century involved fermenting food, cooking it slowly in fat (a *confit*) or crystallising it in sugar. But these diverse treatments only worked with certain types of food, and some had other drawbacks, such as radically altering the taste or removing much of the nutritional value. Vegetables out of season were simply unknown. Deprived of fresh produce, sailors on long voyages often contracted scurvy, a serious, even fatal, illness brought on by a deficiency of vitamin C.

Before Pasteur

Nicolas Appert, the son of an innkeeper, was born in the Champagne region in 1749. Aged 24, he travelled to Germany to complete his apprenticeship as a confectioner in the kitchens of royal households, and it was there that he began to appreciate the importance of keeping food fresh. Back in France, he experimented with conserving food and in 1795 came up with a practical method that involved sealing food hermetically in bottles, then subjecting it to a long rolling boil in a water bath, which sterilised the contents. Appert used empty champagne bottles, as at the time these were the only containers strong enough to withstand the boiling process. We now know that the heat of boiling destroys the micro-organisms

Cannery row
Tin cans for food, from the early 20th century. Cans made of tinplate gradually supplanted the glass bottles used by Appert.

A CLOSELY GUARDED SECRET

Tinplate, now universally used for canning food, dates back to the 12th century, when the process of covering sheets of mild steel with a thin layer of tin was a skilled and closely guarded industrial secret known only to the metalworkers of Saxony and Bohemia. In the 17th century Jean-Baptiste Colbert, minister of finance to Louis XIV, persuaded some German tinplating experts to set up Europe's first tinplate factory at Beaumont-la-Ferrière. With the advent of the Industrial Revolution, Britain became the world leader in the manufacture of tinplate. Famously, the British explorer Captain Scott took tins of Heinz food, including steamed puddings, on his ill-fated Antarctic expedition of 1910-11. They were opened 47 years after his death and found to be perfectly edible.

and enzymes that cause food to go off, and that hermetic sealing prevents recontamination, but it would be another 69 years before Louis Pasteur established the scientific basis behind Appert's technique. Foodstuffs treated in this way kept for several months, even years, while retaining many of their vitamins – a fact not appreciated until 1897.

A head for business

Appert was already a familiar figure in Paris by the time he invented his preserving process. In 1784, on his return to the French capital, he opened the 'Renown' confectioners on rue des Lombards, which soon lived up to its name. Appert had a good head for business and expanded into wholesale confectionary, distributing his fresh sweets throughout the country. Meanwhile, his 'conserves', as he called them, enjoyed even greater success. In 1802 he bought an estate at Massy, just outside Paris, and founded the world's first bottling factory there, employing up to 50 people. The buildings were surrounded by a vast kitchen garden – Appert insisted on bottling produce that had just been picked, so as to preserve as much freshness as possible.

Appert sold his bottles of conserves from his grocery on the rue Saint-Honoré in Paris and he also posted them far and wide. They were not cheap – a bottle cost the equivalent of a day's wages for a worker – but wealthy food lovers throughout Europe vied with one another to be the first to sample vegetables out of season. The famous gourmet Alexandre Grimod de La Reynière even pronounced them superior to the fresh article. Garden peas, which were the first vegetables Appert conserved, were the star of the show, but he was not afraid to diversify his product range, even going so far as to introduce whole prepared meals based around fish in sauce.

Food industry
A print from the mid-1870s shows tins being filled and soldered shut in a cannery. By this stage, food preservation had long since left behind its origins as a cottage industry and developed into a full-blown manufacturing process.

Lifting the lid *The first rather crude can openers did not appear until 50 years after the invention of canned food.*

CAN OPENERS

For a long time, opening a can was by no means straightforward. The first tins were made of very thick steel and were extremely resilient. There was no purpose-made implement for getting into them, so people resorted to hammers and chisels. Things improved somewhat with the appearance of cans made from thinner, unsoldered, tinplate. But an effective solution did not come until 1858, when American Ezra Warner invented the can opener. The gadget was immediately adopted by the US (Union) Army, which used it extensively during the Civil War. In 1866, an inventor named J Osterhoudt patented a tin can fitted with an integral key opener, which is still widely used on cans of sardines, anchovies and corned beef. The familiar modern-style can opener, with a wingnut turning mechanism and a cutting wheel, was invented by William Lyman of Connecticut and first appeared in 1870. The Star Can company of California improved on his design in 1925, adding a serrated wheel to grip the can more firmly. The first electric can opener came on the market in 1931.

Rise and fall

Appert appeared to strike gold in 1810, when the French Navy awarded him a contract to supply preserved food. But there was a condition: that he make known his method of conservation. Accordingly, the following year he published *The Art of Preserving Animal and Vegetable Substances for Several Years*, which was immediately translated into several languages. The contract earned him the sum of 12,000 francs, and obliged him to adopt a whole new scale of mass production. Yet no

MAN OF MANY PARTS

Although the food preserving process remains his most famous invention, Nicolas Appert spent his whole life improving other techniques and pieces of technology. For example, he found that heating milk to 70°C did not significantly impair its flavour but enabled it to be kept for several weeks. This process, known as 'appertisation', anticipated pasteurisation. But Appert's methods had limitations: because a water bath could not get above 100°C, the operation did not kill all germs. Meat, in particular, remained susceptible to the bacterium that caused botulism, a fatal form of food poisoning. So Appert took Denis Papin's steam digester – a forerunner of the pressure cooker, invented in 1679, in which steam pressure produced a temperature of 115°C – and refined it into the autoclave, a high-pressure vessel which he used to sterilise his conserve bottles.

Under pressure *The original autoclave cooker designed by Appert.*

AROUND THE WORLD WITHOUT SCURVY

On 17 September, 1817, the French naval corvette *L'Uranie* left Toulon under the command of Captain Louis Claude de Saulces de Freycinet. She carried in her holds a large quantity of Appert conserves – now contained in tinplate cans rather than bottles. This was to be a long voyage; her mission, ordered by Louis XVIII, was to circumnavigate the globe 'with a view to plotting the precise shape of the Earth, studying its magnetic force, and gathering all manner of natural history specimens that might contribute to the advancement of science'. On 15 February, 1820, *L'Uranie* foundered off the Falklands. Her crew all escaped unharmed and completed their voyage on a vessel that they purchased on the islands. They finally returned to Le Havre on 13 November, 1820, with not one of the men having contracted scurvy.

sooner had he made his fortune than it was snatched away. A series of French defeats and the British naval blockade of European ports strangled the market for his conserves. In 1814–15, the factory at Massy was pillaged first by the Prussians and then by the British. Appert never recovered financially and died in penury at the age of 91.

From the jar to the tin

Following the publication of Appert's work, preserving factories sprang up all over Europe. In 1810 Pierre Durand, a French émigré in London, licensed the idea of replacing heavy, fragile bottles with tinplate cans, which were not only lighter and more durable but could also hold more (up to 10 kilograms). Bryan Dorkin and John Hall acquired the patent and in 1812 set up a factory dedicated to supplying canned goods to the British army. In 1846, Henry Evans invented a machine that could turn out 60 cans an hour, ten times faster than any previous process. The canning industry was now in full swing, freeing domestic cooks from using only seasonal produce and changing army rations beyond all recognition.

In the 1860s, Louis Pasteur invented a preserving process that sterilised food at lower temperatures. Although this meant that food could not be kept for as long, it did give it a better flavour. Appertisation, pasteurisation, and later refrigeration (introduced from the 1870s) brought consumers a choice of foods unimaginable a century earlier.

Military rations
A tin decorated with the head of Queen Victoria, issued to troops fighting in South Africa in the Second Boer War (1899–1902). Tins had other uses: soldiers often banged on empty cans to warn of an enemy attack.

Famous icon
The Campbell's Soup can is one of the best-known tinned products in the world, thanks to being featured on a series of screen-prints by the American pop artist Andy Warhol in 1966.

Homeopathy 1796

Samuel Hahnemann, a leading German scholar and medic, spent much of his time translating important scientific works from abroad. In 1790, while reading *A Treatise on the Materia Medica* by Scottish physician William Cullen, he found himself questioning Cullen's assertion that Peruvian cinchona bark (quinine), commonly prescribed for people suffering from intermittent fever (malaria), acted through its 'tonic effect on the stomach'. Hahnemann took a different view: after testing quinine on himself, he reached the radical conclusion that 'Peruvian bark ... works because it can produce symptoms similar to those of intermittent fever in healthy people'.

Alternative pills
Like conventional pharmaceuticals, homeopathic remedies can be administered in the form of tablets (right) or drops in solution.

Curing like with like

Armed with this insight, in 1796 Hahnemann formulated his principle of similitude, which became the basis of an alternative medicine he called homeopathy (from Greek *homoeos*, meaning 'similar', and *pathos*, 'illness'). Curing an illness involved administering a remedy that would induce in a healthy patient 'a morbid state as similar as possible to the case of disease before us'.

IN MINUTE DOSES

Homeopathic cures are prepared by making a mother tincture – a strong solution of an animal, plant or mineral substance – to which is added a base (water or alcohol) in dilution steps of 1 to 100. The Hahnemann centesimal scale (1 CH, 2 CH to 100 CH) indicates the potency of the final dilution.

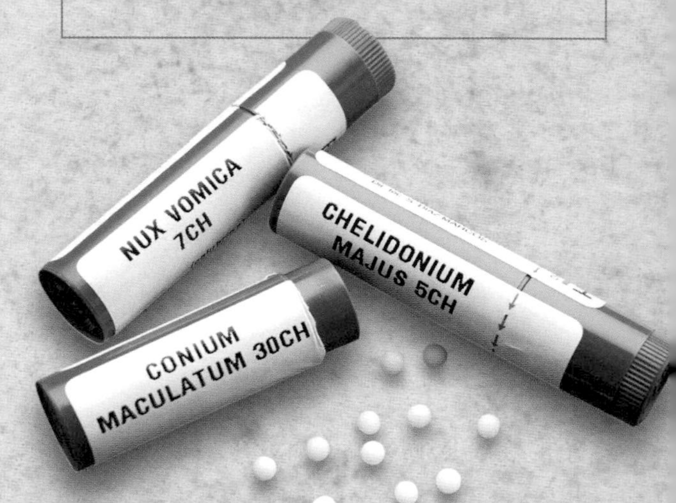

A growing interest in the effects of drugs led him to reject the prevalent empirical approach among doctors, who treated disease with drugs that produced opposite effects to those of an illness. Hahnemann advocated listening carefully to the patient before applying remedies. In his major work, the *Organon of Rational Healing* (1810), he detailed the way in which he prepared weak dilutions of drugs. By the time of his death, in 1843, he had tested numerous chemical compounds and mapped out a whole typology of different people's susceptibility to chronic diseases and appropriate remedies.

Right from the outset, Hahnemann's therapeutic method excited controversy. He was bitterly attacked by the German medical establishment and others. But homeopathy won many followers, and before long advocates of alternative medicine were spreading Hahnemann's remedies.

Historic kit
A portable homeopathic medicine chest from the late 19th century containing 34 vials of drugs and tinctures.

SMALLPOX VACCINE – 1796

Immunity through infection

Smallpox was a killer, but a folk remedy claimed that you could avoid contracting the feared disease by infecting yourself with cowpox. British doctor Edward Jenner decided to investigate the phenomenon and his work led to the world's first truly preventive cure – vaccination. It was the first victory in the long-running fight to combat infectious diseases. Less then 200 years after Jenner's discovery, smallpox was effectively eradicated.

Momentous discovery
A glazed French earthenware plate from the late 18th century celebrating the origin of vaccination.

During the 18th century, Britain was ravaged by a succession of smallpox epidemics. The disease killed thousands every year – 10 per cent of all deaths in London in the century were ascribed to it – and disfigured even more with the permanent scars that it left behind. Edward Jenner, a young country doctor from Gloucestershire, was all too aware of the effects of this terrible scourge on his home village of Berkeley, where he had opened a practice in 1773. Yet a piece of local folk wisdom claimed that all the milkmaids who caught the cowpox virus remained unaffected by smallpox. Cowpox manifested itself in pustules on the cows' udders. Anyone who survived smallpox was immune to the disease thereafter, and Jenner was keen to find out whether cowpox might have the same effect.

Twenty years of study

From 1775 onwards, Jenner undertook extensive research into the symptoms of cowpox and confirmed that the disease could be transmitted from cows to humans, and that people who caught it once could not contract it again. The effects of cowpox on humans appeared relatively benign, with the blisters that swelled up on the hands disappearing after a few days. He was also able to corroborate the fact that infection by cowpox made people immune to smallpox. He was

AN ANCIENT TECHNIQUE

Variolisation – preventive inoculation with a mild form of smallpox– is thought to have been practised in ancient China and India from c200 BC or even earlier. Traders on the Silk Road introduced it to the West. By the 1700s, it was being widely practised in Constantinople, where Lady Mary Wortley Montagu (left) was wife of the British Ambassador. Lady Montagu had her own children immunised, then introduced the practice to England on her return in 1721. Thereafter, smallpox inoculation became widespread in Europe, due in large part to Theodore Tronchin, a Swiss physician who worked in many aristocratic households.

Pioneer in deeds and words
Lady Mary Wortley Montagu (1689–1762), wife of the British ambassador to the Ottoman Empire, shown in a portrait by Jonathan Richardson (left). Her Letters from Turkey, *published in 1763, are regarded as the earliest example of women's travel writing.*

SCOURGE OF EAST AND WEST

Smallpox (variola virus) is thought to have appeared in Mongolia *c*1000 BC. The first known descriptions of its symptoms – severe fever and pustules on the skin – come from 4th century AD China. By the 6th century, it had spread to the Middle East and Turkey. The first cases in western Europe were recorded in the 16th century, and from there it was taken to America. Two centuries later, the disease was killing up to 400,000 Europeans annually.

A medical first
James Phipps being inoculated against smallpox by Edward Jenner in 1796, recorded on canvas (below) by the French artist Gaston Melingue.

now determined to find a way of immunising as many people as possible against the dreadful disease.

As a preventive measure, he began inoculating healthy subjects with pus taken from others infected with a mild form of smallpox. Although it was reputed to protect people from later infection, the method was unreliable; it was not always effective and it could even prove downright dangerous, producing severe adverse reactions and even fatalities. But would vaccination with cowpox produce more controllable results? His former tutor, the famous London surgeon John Hunter, wrote to encourage him: 'Don't think, try; be patient, and accurate.' Jenner stuck to his task for many years, diligently observing the symptoms of cowpox and carrying out various experiments. He found out, for instance, that cowherds who had contracted cowpox did not react adversely to variolisation (inoculation with smallpox), but that the degree of protection it afforded depended upon how advanced the cowpox was. He also tried out different methods of vaccination.

Inoculation instrument
One of the original needles used by Edward Jenner to inoculate patients with the cowpox virus.

Conclusive proof

In May 1796, Jenner diagnosed cowpox in a dairymaid by the name of Sarah Nelmes. Extracting some pus from the blisters on her hands, he used it some time later to inoculate eight-year-old James Phipps, without any ill effects. This young boy, the first recorded person to receive a cowpox vaccine, was now immunised against smallpox. Jenner went on to repeat the procedure 30 times or more, proving conclusively that vaccination was more effective and safer than variolisation. He published his findings in 1798, in a work entitled *An Inquiry into the Causes and Effects of the Variolae Vaccinae*.

Initially, Jenner's work was regarded with suspicion, but ultimately he was to enjoy stunning success. By 1800, his vaccination had spread to eastern Europe, Russia Scandinavia and the Mediterranean region, and to Latin America by 1803. For the first time, leaders of nations incorporated public health into their political agendas. Industrialisation, major public works and military campaigns all made huge demands on manpower, and immunity from smallpox helped to offset the mortality rate. Napoleon ordered his troops to be vaccinated in 1805. A British Act of Parliament of 1853 made smallpox vaccination compulsory for the entire population. As the 19th century ended, smallpox was on the wane in Europe and was largely confined thereafter to tropical areas of the world.

Combating infectious diseases

Jenner's discovery marked the start of the worldwide campaign against epidemic diseases. Later, the work of Louis Pasteur and his successors would identify numerous bacterial and viral agents and throw more light on the process of immunisation. As a result, vaccines were introduced against other diseases: rabies (1885), typhoid fever (1896), tuberculosis,

whooping cough, diphtheria and tetanus (all in the 1920s) and yellow fever (1937). Smallpox vaccination continues to serve as a model in the fight against epidemics. From 1917 onwards, improved methods of conserving the vaccine, by freeze-drying it and sealing in it glass vials, contributed to the spread of mass immunisation programmes even to the world's remotest places.

The last case of smallpox was reported in Somalia in 1977. Three years later the World Health Organization declared that it was the first disease to have been officially eradicated. But the fight against infectious and contagious diseases is far from over. While some diseases are in retreat, others have emerged, such as AIDS and the Ebola virus. Medical researchers are engaged in a constant struggle to find antidotes and cures.

Matron knows best
Spanish schoolchildren stand in line with sleeves rolled up, ready to be vaccinated against smallpox in 1961.

Smallpox at close quarters
The smallpox virus (above) as seen under an electron microscope. The virus causes a highly contagious disease, with symptoms of high fever and a rash of swellings on the skin that develop into pustules.

Sailing in the name of science

From the mid-18th century onwards, new horizons began to open up for scientific ocean-going exploration. Unlike in the 1500s, the great navigators of this new generation – notably Louis-Antoine de Bougainville and James Cook – were no longer motivated by the lure of gold and military conquest. Scholars and skilled seafarers joined forces to chart previously unknown lands.

Position finder *Made by Scottish instrument maker James Ferguson in 1767, this brass armillary sphere (below) was used to fix the position of a ship. Its rings show the movements of stars in relation to the Earth.*

Maritime exploration in the 16th century was in many regards a clandestine and haphazard business. Navigators and their royal patrons met in secret behind closed doors to plan expeditions, and colluded in throwing potential rivals off the scent. The Portuguese court, for instance, was all too eager to present foreign ambassadors with lavishly produced maps of the world – peppered with false or at best partial data. In the absence of hard information, long-distance seafarers made halting progress relying largely on chance.

Towards the end of the 17th century, the first royal observatories dedicated to improving navigation were founded at Greenwich and Paris. For the scientific community that was fast developing at the time, imprecise accounts by sailors were no longer acceptable. Rather, as exploration became a state-organised enterprise, the men of science were called upon to pinpoint new areas of interest and bring intellectual rigour to bear on new discoveries. Furthermore, the up-and-coming generation of naval officers was given a thorough grounding in the sciences, to the extent that many of them became respected scholars in their own right.

Tales of exotic lands

The great age of discovery may be said to have begun on 15 December, 1766, when the frigate *Boudeuse* left the French west coast port of Brest in a stiff breeze to begin a circumnavigation of the globe. Her captain, Louis-Antoine de Bougainville, had been commissioned by Louis XV to undertake the voyage. He was accompanied by the naturalist Philibert Commerson and the astronomer Pierre-Antoine Véron. After linking up with

Land ahoy! *A painting by Ambroise Louis Garneray of 1768 depicts Bougainville and his crew (below) discovering the Louisiade Archipelago, a string of volcanic islands east of Papua New Guinea.*

A MAN OF HIS WORD

The son of a notary, Louis-Antoine de Bougainville (1729–1811) studied science, mathematics, literature and law. After working as a lawyer and serving in the army, he was appointed secretary to the French ambassador to London in 1754. During the French and Indian Wars in Canada, he served under General Montcalm, before joining the navy. His most famous voyage took him to Tahiti, where a young Polynesian named Aotourou asked leave to accompany him back to Paris; Bougainville agreed, promising to return him to Tahiti. Aotourou duly caused a sensation in France, then Bougainville spent much of his personal fortune trying to keep his word. Sadly, the Tahitian died on the return journey. A flowering plant was discovered during a stopover in Brazil and named Bougainvillea by the expedition's botanist, Philibert Commerson.

Tahitian beauty *Poetua, daughter of chief Oreo of Raiatea in the Society Islands. On his third and final voyage to the South Seas on his ship* Resolution *in 1776–9, Captain James Cook took Oreo and his family hostage to secure the return of crew members who had mutinied, opting for a life of ease on the island.*

the supply ship *Étoile* off Brazil, the expedition sailed through the Straits of Magellan and explored the Tuamotou Archipelago before dropping anchor off Tahiti in April 1768. Bougainville thought that he was the first European to set foot on the island, but in fact the British seafarer Samuel Willis, captain of HMS *Dolphin*, had landed there almost a year before. The French scientific team set about studying not just the island's flora and fauna, but also its inhabitants. Bougainville noted: 'The native women here are characterised by a gentle languor, and their chief concern seems to be to please the menfolk.'

On his return to Saint-Malo in 1769, Bougainville recounted the story of his expedition in *Voyage Round the World,* and the work became an instant hit. Reports of the innocence and sexual license of the Tahitians fed into the myth of the 'noble savage' promulgated by the Enlightenment philosopher Jean-Jacques Rousseau.

Yet these rather dubious reasons for the success of Bougainville's published work obscured the truly important scientific insights gained by his expedition. Véron, for example, used the lunar-distance method to determine longitude, becoming the first person to fix the position of the Philippines. In addition, many new islands were discovered, while ethnographic accounts yielded new insights into the customs of the indigenous peoples of Oceania. The significance of the voyage was not lost on the writer Denis Diderot, who wrote *Supplement to the Voyage of Bougainville* in 1772.

Terra Incognita Australis

British seafarers were not to be outdone. On 26 August, 1768, James Cook embarked on the first of his hugely successful voyages of

DES VAISSEAUX DU ROY
LA BOUDEUSE ET L'ETOILE AUTOUR DU MONDE
AMERIQUE SEPTENTRIONALE
TROPIQUE DU CANCER
LIGNE EQUINOCTIALE
TROPIQUE DU CAPRICORNE
MER PACIFIQUE
ROUTE DE L'ENDEAVOUR
COTES DECOUVERTES PAR COOK
ROUTE DE BOUGAINVILLE
Extrait d'une édition des voyages de Bougainville appartenant à un ami du Capitaine Cook. Le tracé des routes des navires est dû à Cook.

Epic voyage *A copy of an 18th-century map showing the route of the* Boudeuse *and the* Étoile *in the Pacific during Bougainville's circumnavigation of the world.*

Faraway land
A coastal landscape on New Zealand's South Island. James Cook reached New Zealand in 1769, on his first Pacific voyage. In the course of plotting an accurate chart of the coastline, Cook discovered that New Zealand comprised two islands. The passage between them now bears his name – Cook Strait.

Crossing the line
On 17 January, 1773, James Cook became the first navigator to cross the Antarctic Circle, at 66° 33' South. The watercolour of the Resolution *below, painted by a crew member, shows men breaking up ice for water.*

exploration under the auspices of the Royal Society. Cook was a brilliant navigator who had trained on coal ships out of Whitby. For his new venture he chose a robust former collier, HMS *Endeavour*. Passing through the Straits of Magellan, he arrived in Tahiti a year after Bougainville. From there, on 3 June, 1769, he observed the transit of Venus across the Sun. Also on board was the naturalist Joseph Banks and botanist Daniel Solander, who carefully catalogued and described the plant and animal species of the island.

The expedition then headed for New Zealand, where it survived an attack by hostile Maori warriors. Moving on to the east coast of Australia, Cook landed at a place he named Botany Bay on 29 April, 1770, and spent several months mapping the coastline. He also

FROM THE CARAVEL TO THE FRIGATE

In the 18th century, ships-of-the-line became the backbone of all major European navies. The high forecastles and sterncastles of old Spanish galleons gradually disappeared as warships took on a more horizontal profile. Ships-of-the-line and the smaller frigates had three masts, raised taller than ever to take larger areas of sail. Smaller sails fore and aft helped with steering. Oak hulls grew longer, thanks to the use of metal fixtures. Sailing ships reached a peak of development in this period and remained in frontline service until the arrival of the first steam-and-sail powered ironclads in the late 1850s.

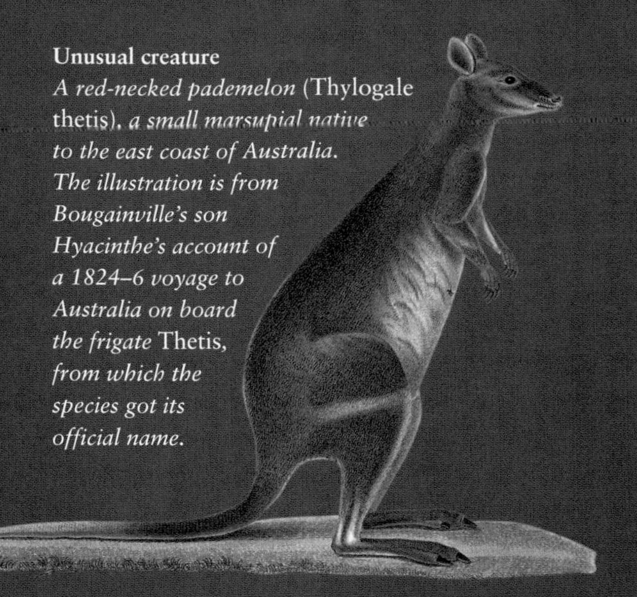

Unusual creature
A red-necked pademelon (Thylogale thetis), *a small marsupial native to the east coast of Australia. The illustration is from Bougainville's son Hyacinthe's account of a 1824–6 voyage to Australia on board the frigate* Thetis, *from which the species got its official name.*

discovered a bizarre animal which 'moves by running and hopping' – the kangaroo. Cook returned to England in June 1771.

He set sail again the following year, venturing as far as the Antarctic Circle. Part of his mission on this second voyage was an attempt to locate the legendary *Terra Incognita Australis*, a vast southen continent generally supposed to exist by the geographers of the Middle Ages. His exploration of the ice floes in the southern ocean lasted three months without yielding any evidence of the fabled continent. Cook concluded that, if it existed at all, the land must be uninhabitable. Abandoning his quest, he set a course first for Tahiti and then on to Easter Island. He noted in his log that the giant statues there 'testify to the level of industry and perseverance shown by the islanders'.

Returning to England, Cook was tasked next with finding a Northwest Passage linking the Atlantic and Pacific oceans. He discovered Hawaii, which he named the Sandwich Islands, in 1778, then proceeded to sail up the northern Pacific coast of America, but thickening ice forced him to turn back. As a result he revisited Hawaii, with fatal results: the great seafarer was killed in a skirmish with Hawaiian islanders in February 1779.

Despite the tragic loss of its captain, the expedition continued to search for a Northwest Passage, but after failing to find one headed back to England. Cook's achievements on his three voyages were many and varied, but among the most important were the superbly detailed maps he left behind. He also instituted a diet rich in vitamin C – a mix of lemon, onion and pickled cabbage – which protected his crews from the dreaded scurvy. His pioneering seamanship opened the South Seas for later explorers.

The death of James Cook
On 14 February, 1779, while visiting Hawaii (the Sandwich Islands), Captain Cook was clubbed and stabbed to death during an altercation with islanders in Kealakekua Bay. His remains were later buried at sea.

NATIONAL HERO

Born in 1728 into a family of Yorkshire farmers, James Cook signed on at an early age to be a ship's apprentice. He showed real passion for the sea and natural leadership qualities, and at the age of 27 his mentor, Captain John Walker of Whitby, offered him command of a collier brig. That same year, 1755, Cook volunteered for the Royal Navy. He was a stickler for discipline and brave in battle, and soon gained promotion. He taught himself maths, astronomy, navigation and trigonometry and was chosen by the Royal Society to command its voyages of exploration. These voyages made Cook a national hero, and his death enshrined him in legend. Over two centuries later, the names of two of his ships – the *Endeavour* and *Discovery* – were adopted by NASA for their Space Shuttles.

Ill-fated expedition

One voyager who attempted to emulate Cook was Jean François de La Pérouse, who set sail from Brest in August 1785 on an expedition sponsored by Louis XVI. At his command were two sturdy former merchant vessels of 500 tons apiece, the *Boussole* and the *Astrolabe*, and a crew of 220 men. Also on board was a sizeable scientific team.

The expedition's aims were to complete the mapping of the world, open up new sea lanes and trade routes and collect new specimens for study. They made steady progress westward, to Brazil and then around Cape Horn, before taking a looping course north around the Pacific, calling at Easter Island, Alaska, Macao, Tonga and Australia. Occasional dispatches hinted at La Pérouse's ambitions: 'At the end of July 1788, I intend to sail between New Guinea and New Holland [Australia] on a different course to that taken by the *Endeavour*, if such a route be at all possible…'. Then came silence: the expedition simply vanished off the face of the Earth.

The mystery remained unsolved for many years. The last words of Louis XVI, just before being guillotined, were reputedly: 'Is there any news of Monsieur de La Pérouse?' In 1827, some wreckage that appeared to come from his ships was found on Vanikoro in the Solomon Islands. This has since been confirmed as the wreck of the *Boussole*.

Unknown species

The fourth key explorer of this period was Alexander von Humboldt, a scientist rather than seafarer. The son of a wealthy Prussian family, he travelled widely throughout Europe,

Grand plans
A painting of 1817 by Nicolas Monsiau (below) shows Louis XVI (seated centre) giving instructions to La Pérouse (left) before his ill-fated voyage. On the right, behind the King, is the French naval minister, the Marquis de Castries.

Precious specimens
A page from the sketchbook of the botanist Gaspard Duche de Vancy, a member of La Pérouse's expedition. These detailed cutaway drawings show a herbarium, used for transporting bush and shrub specimens back to Europe.

ARISTOCRAT ADVENTURER

Jean-François de Galaup, Comte de la Pérouse (1741–88) joined the Marines in France at the age of 15 and sailed widely on the world's oceans before being given command of a series of French ships-of-the-line. The courage he showed in several naval engagements, together with his erudition and skill as a diplomat, earned him command of the most ambitious expedition ever mounted by Royalist France. La Pérouse's aim was to surpass the achievements of his most illustrious predecessor, James Cook, but his two ships vanished. Proof of their fate emerged only in 1964.

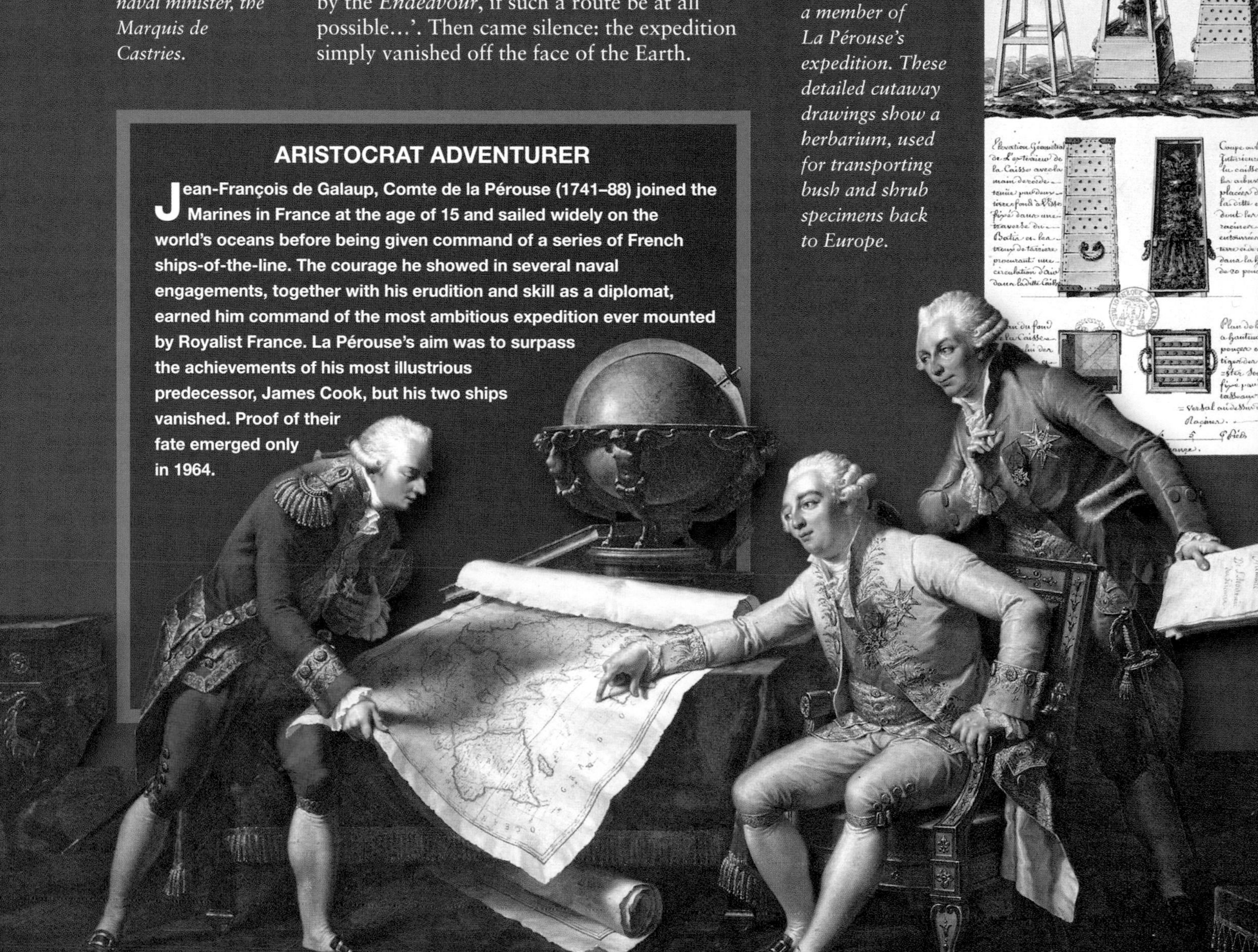

Asia and the Americas in search of new plant and animal species. Accompanied by the French botanist Aimé Bonpland, Humboldt journeyed up the Orinoco River through Venezuela and Colombia. He then set out to climb Mount Chimborazo, a volcano that was thought at the time to be the highest peak in the world. The two intrepid explorers were forced to turn back just 500 metres from the summit. On their return to Europe in 1804, they brought back sheaves of sketches, detailed maps and a herbarium containing 5,800 plant species, 3,600 of them unknown.

It was through the efforts of the explorers of this golden age of discovery that the world began to be accurately mapped. For good and ill, their work paved the way for the large-scale colonial expeditions of the 19th century, leading on the one hand to greater scientific understanding of the natural world, on the other to systematic exploitation of its resources and peoples.

PRUSSIAN POLYMATH

Alexander von Humboldt (1769–1859) was born into a Prussian aristocratic family and educated by the foremost tutors of the age. He took a keen interest in many branches of the natural and social sciences – geography, geology, anthropology, climatology, economics – and after exploring areas of the Rhine, as well as Italy and Switzerland, he used his fortune to fund a five-year expedition to Latin America from 1799 to 1804. He was a gifted historian, artist, painter and cartographer. By the time of his death, he had completed 34 volumes of his monumental *Journey to the Equinoctial Regions of the New Continent*, along with several atlases. Humboldt voiced his opposition to slavery and colonialism in letters to world leaders, including US President Thomas Jefferson.

Under the volcano
Alexander von Humboldt and Aimé Bonpland in the Tapia Valley below Mount Chimborazo, in an oil painting of 1810 by Friedrich Georg Weitsch.

THE BOARD OF LONGITUDE

In 1707, a miscalculation of his fleet's position by Admiral Sir Cloudesley Shovell led to his flagship, HMS *Association*, and several other vessels running aground off the Scilly Islands. The Admiral and 1,400 men perished in the disaster. In response, in 1714 the British government set up the Commissioners for the Discovery of Longitude at Sea, commonly called the Longitude Board. In the 114 years of its existence, the Board's most important act was to commission new, accurate marine chronometers from the instrument-maker John Harrison and his son William. Harrison laboured on his superbly reliable timepieces from 1730 to his death in 1778, but it was only in 1773, after the personal intervention of George III, that he gained his just recognition by the Board. Part of the problem was Nevil Maskelyne, the Astronomer Royal and a member of the Board, whose own 'lunar-distance' method was in direct competition with the Harrisons' chronometers. James Cook took a Harrison H4 chronometer with him on his second and third Pacific voyages, and reported finding it more accurate than the lunar-distance method that he used on his first voyage.

Mapping the world
'A Chart of the Southern Hemisphere' published in James Cook's account of his voyages. The map shows the momentous journeys undertaken by the great navigators of earlier eras, while around the margins Cook noted the latitude and longitude of all the islands he discovered in the South Pacific.

Lithography 1796

In late-18th-century Munich, a struggling actor and playwright, Alois Senefelder, hit upon the money-saving ruse of printing his own plays. He did this by engraving the text onto thin copper sheets using a fine engraver's cutting tool – a graver – and acid. The process was arduous and painstaking: Senefelder had to use mirror-writing and he soon realised that not only was copper expensive, but it was also hard to work with. So he began looking for a cheaper and more malleable material. He chose Kelheim limestone, a smooth, fine-grained stone from the part of Bavaria where he grew up that was used to make flagstones.

The future of printing
An original Senefelder lithographic printing press, dating from 1793 (below).

A chance discovery

One day in 1796, just as he was about to polish a piece with fine sand, his mother asked him to work out what they owed the laundress. Having no paper to hand, Senefelder jotted down the figures on the stone, using a greasy ink made of wax, soap and lamp-black (carbon). Later, he decided to clean the stone plate with the same acid that he had used for his copper engraving. The acid ate away at the stone but left the area protected by the ink standing out in relief. Simply by inking up the plate again, he was able to use it for block printing.

Breakthrough in stone
An engraving of 1843 depicts Alois Senefelder working out the family laundry bill on a piece of limestone, which inspired his discovery of a new method of printing.

Experimenting with a fresh piece of limestone, Senefelder realised that ink adhered to areas that he wrote on with his waxy crayon, but not to any part of the stone wetted with water. And thus was born the idea of lithographic printing – that is, printing without engraving. The term 'lithography' comes from the Greek *lithos*, meaning 'stone', and *graphein*, meaning 'to write'.

Before Senefelder's discovery, the only method of printing was letterpress, which involved inking the raised surfaces of etched, engraved or cast characters and pressing them onto paper. In contrast, lithography was based on the mutual repulsion of oil and water. Words or images were applied to the plate in crayon, then the surface of the plate was treated with a mixture of gum arabic and nitric acid, which was washed off after a while. The areas of the plate without waxy crayon were attacked by the acid and became water-absorbent, while those protected by crayon remained water-repellent. All that remained was to roll ink over the plate, lay a sheet of

Boom time
The heyday of lithography as an art form was in the 1890s. This poster by the illustrator Fernand-Louis Gottlob (left) is for the second exhibition of works by lithographic artists in Paris in January 1899.

paper on top and apply pressure with a block to obtain a printed page. Senefelder solved the problem of back-to-front writing with tracing paper. He wrote as normal on the paper, then inverted this onto the printing plate. The final impression was inverted again, and so appeared the right way round.

Fame and fortune

Senefelder took out Europe-wide patents on his invention, and published a detailed account of it in 1818, which appeared in English the next year under the title *A Complete Course of Lithography*. At first, though, lithography was regarded as crude in comparison with letterpress and struggled to make headway. Another difficulty was colour printing; initially, black and white prints were simply coloured in by hand. But in 1838, a printmaker from Alsace named Godefroy Engelmann revived a four-colour printing process originally devised by a German engraver, Jacob Christoph Le Blon. By combining the primary colours red, blue and yellow with black, he found he could reproduce every shade. He added a device that aligned the paper precisely during printing, a vital tool to avoid colour-bleed. Engelmann's process became known as chromolithography.

Aluminium sheets later replaced the stone plates, an innovation that went hand-in-hand with the ability to capture a whole page – text and images – photographically, then transfer the negative to the litho plate. By mounting the plate and paper on cylinders and adding a rubber-covered drum (the blanket cylinder) to transfer ink from plate to paper, offset lithography was born. This revolutionised printing, especially newspaper production, from the late 19th century onwards.

Portrait of a lady
A lithograph and watercolour poster of the actress Sarah Bernhardt made by Alphonse Mucha in 1896 (below). In the 19th century, chromolithography became the main printing method for reproducing works of art, and for colour illustrations in books and magazines.

PRINTMAKING AS AN ART FORM

The instant visual impact of lithography soon drew the attention of artists. From the mid-19th century, painters such as Géricault, Delacroix, Daumier and Goya began making black-and-white lithographs. Later, Impressionists like Manet, Degas and Renoir were keen to explore chromolithography and the nuances of colour it could convey. But lithography's heyday really came in the 1890s with the pioneers of poster art, notably Henri Toulouse-Lautrec and Alphonse Mucha. Some 20th-century artists dabbled in lithographic printmaking, but it was gradually abandoned as they reverted to more lucrative one-off canvases. Thereafter, lithography largely became the preserve of graphic illustrators in the advertising industry.

MACHINE TOOLS – 1797

A new age of machines

A key aspect of the Industrial Revolution was the steady replacement of craftsmanship with mass-production by machines. Fundamental to this change was the invention of machine tools that brought accuracy and consistency to manufacturing. The father of this development was Henry Maudslay, who in 1797 invented the screw-cutting lathe.

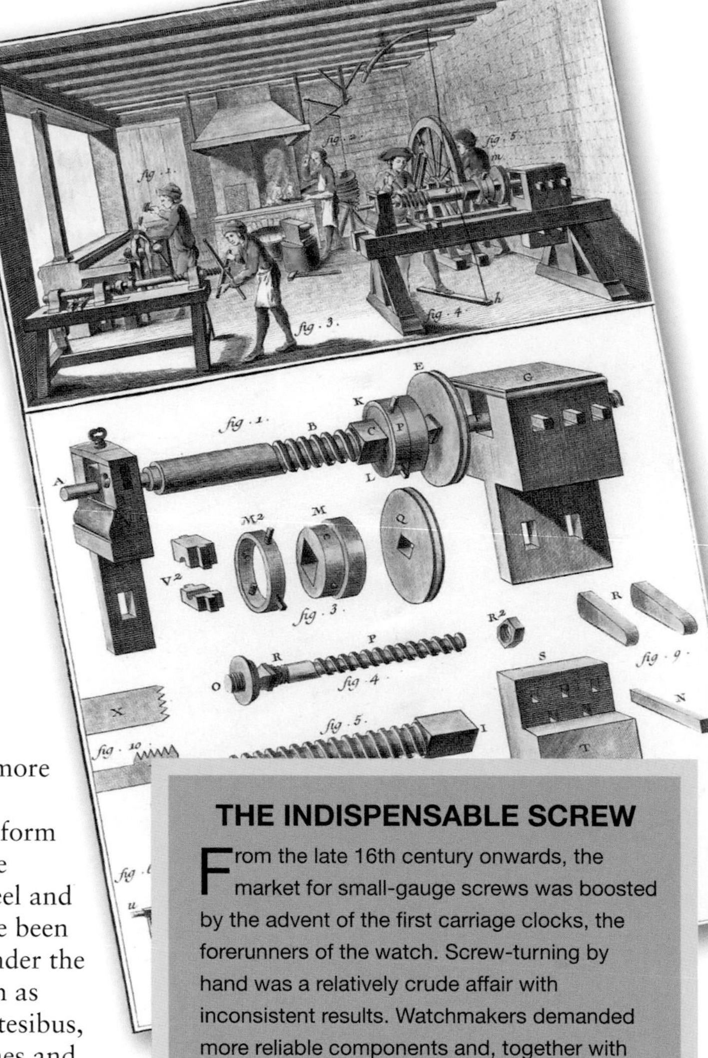

Art of the toolmaker
A plate from Diderot and d'Alembert's renowned French encyclopedia shows a toolmaker's workshop and some of the machinery used.

Precision tool
Vaucanson's slide-rest lathe (below) was completed in 1750–1. The device was used to make large, cylindrical smoothing irons for finishing watered silk fabrics.

The impulse behind the invention of machine tools was threefold. First, to make certain manufacturing processes more simple and precise. Second, to augment human muscle-power. And third, to perform tasks that were impossible by hand. The earliest such tools were the potter's wheel and the wood lathe, examples of which have been found dating back several millennia. Under the influence of ancient Greek scholars such as Archimedes, Hero of Alexandria and Ctesibus, engineers invented elaborate siege engines and lifting gear, which used levers, pulleys and gears to move heavy weights with little effort.

It was only in the Renaissance that people began to grasp the full potential of machines and the possibilities for different applications. This is apparent in many of the sketches of Leonardo da Vinci, as well as in the works of other engineers of the Renaissance and early Modern periods, such as Gerolamo Cardano of Milan (1501–76) and the Frenchmen Salomon de Caus (1576–1626) and Gilles de Roberval (1602–75).

THE INDISPENSABLE SCREW

From the late 16th century onwards, the market for small-gauge screws was boosted by the advent of the first carriage clocks, the forerunners of the watch. Screw-turning by hand was a relatively crude affair with inconsistent results. Watchmakers demanded more reliable components and, together with armourers and locksmiths, were instrumental in the development of machine tools.

A slow process

Little by little, machines transformed the world of work. Foot pedals left the operator's hands free for other tasks. Cranks, flywheels and belt drives gave machines their characteristic rhythm and continuous movement. Machines quickly diversified – different types of lathe, for instance, were developed for turning screw threads, boring cannons, cutting gears and polishing. At first they remained fairly crude, with major components still made of wood. Up to the 17th century, the demand for machine tools was negligible, but with the dawning of the age of scientific and maritime discovery, that situation changed radically. The Baroque style prevalent in the period also created a vogue for small, highly crafted *objets d'art*. As a result, tools became more intricate and precise, primarily at the instigation of watchmakers and opticians. Metalworking skills in brass and other materials also improved markedly. By the mid-1700s, most of the processes involved in machining metal had been mastered: turning, drilling, cutting, milling, shaving, boring and hole-punching. Only the final finishing of metal surfaces was still done by hand with files and chisels.

The impetus behind transforming machines into genuine precision tools came from a number of different innovators in the field.

In around 1750 Jacques de Vaucanson, a French manufacturer famous for his automatons, built the first all-metal slide-rest lathe. This invention could handle long, heavy pieces of metal, greatly reducing the effort and skill required for a worker to complete the same task. Vaucanson's heavy industrial lathe incorporated a sliding tool carriage that was advanced by a long screw. He used it to manufacture parts for the textile industry. Another innovator contemporary with Vaucanson was his compatriot Nicolas Focq, a master locksmith who in 1751 invented the first machine for planing metal.

From water to steam

The technological revolution that swept Britain from the early 18th century onwards created an ideal seedbed for the development of machine tools. Demand for manufactured goods increased exponentially, and the ready availability of materials like cast iron encouraged mechanisation. In 1775 the Cumbrian engineer and industrialist John Wilkinson patented a new kind of boring machine driven by water power via a mill wheel.

Wilkinson's precision instrument was highly effective in producing safer cannons. Yet its main claim to fame was in boring the cylinders for James Watt's first steam engines. The great accuracy of its action ensured that the cylinders it turned out were airtight, an essential quality for the success of Watt's great enterprise. Watt's business partner Matthew Boulton wrote glowingly in 1776: 'Mr Wilkinson has bored numerous cylinders for us, almost without error; on those measuring 50 inches [127cm] in internal diameter, for example, the variation in width is no greater than the thickness of an old shilling down the entire length of the barrel.' The more accurate the initial cylinder boring, the less fuel steam engines used because they worked far more efficiently. The pressure was clearly on for machining to become ever more precise.

Such exacting standards led to the appearance in 1797 of the first true

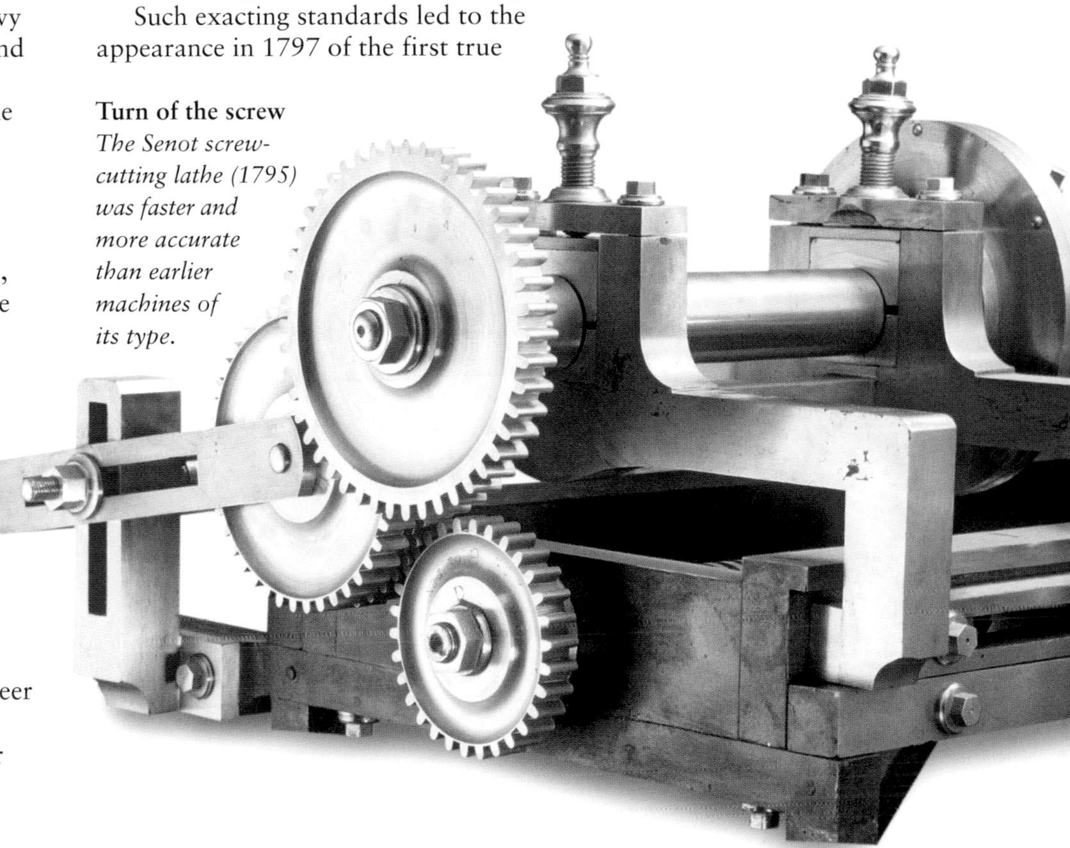

Turn of the screw
The Senot screw-cutting lathe (1795) was faster and more accurate than earlier machines of its type.

PRECISION ENGINEERING

Yorkshireman Jesse Ramsden (1735–1800) was an apprentice in the textile industry before turning to instrument making. His speciality was mathematical and optical instruments. In 1773 he created one of the first 'dividing engines', a device for marking very precise graduations on circular measuring instruments such as sextants, theodolites, armillary spheres and the setting circles of astronomical telescopes. Ramsden's work marked the beginning of precision engineering.

Wilkinson's cylinder boring machine
The motive power for this groundbreaking machine (below) was provided by the mill wheel at its centre. Later versions of Wilkinson's machine were powered by the very steam engines that it had helped to create.

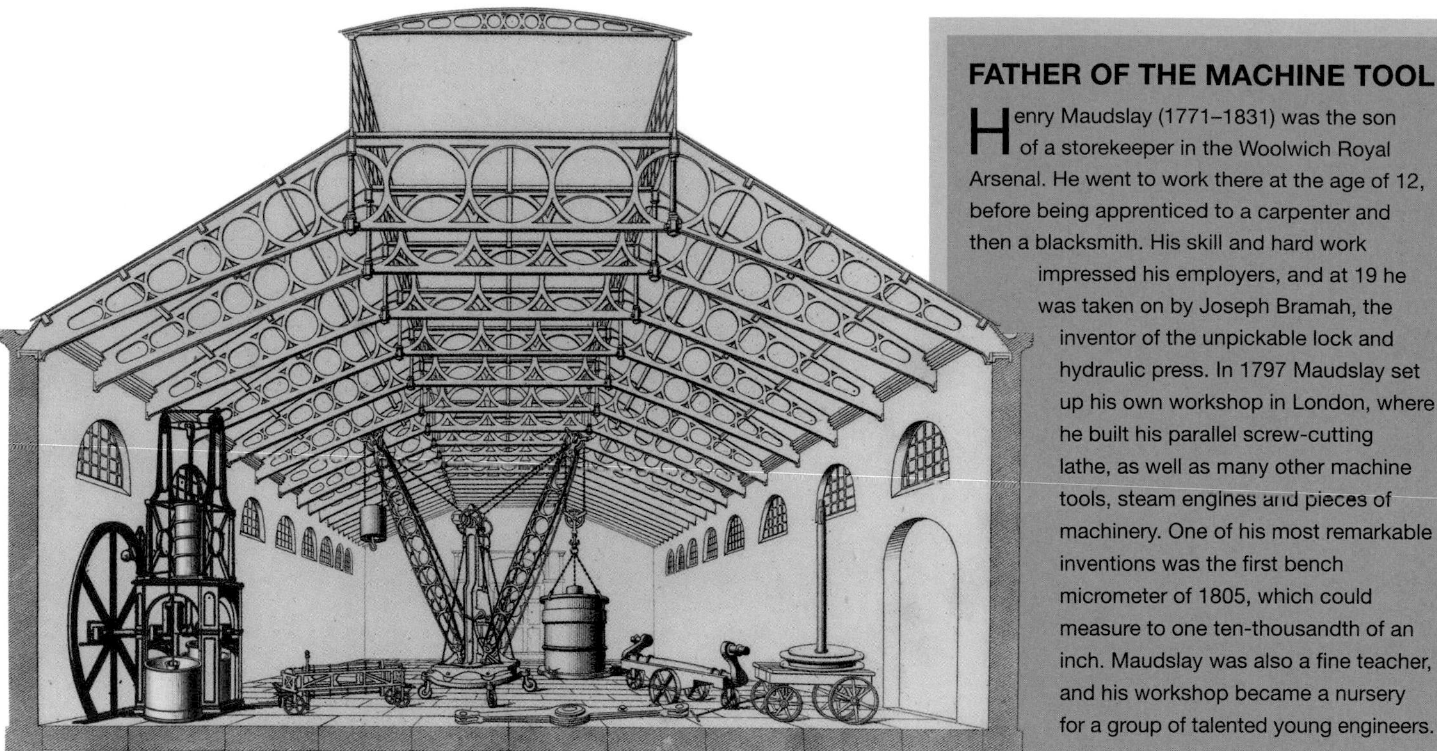

FATHER OF THE MACHINE TOOL

Henry Maudslay (1771–1831) was the son of a storekeeper in the Woolwich Royal Arsenal. He went to work there at the age of 12, before being apprenticed to a carpenter and then a blacksmith. His skill and hard work impressed his employers, and at 19 he was taken on by Joseph Bramah, the inventor of the unpickable lock and hydraulic press. In 1797 Maudslay set up his own workshop in London, where he built his parallel screw-cutting lathe, as well as many other machine tools, steam engines and pieces of machinery. One of his most remarkable inventions was the first bench micrometer of 1805, which could measure to one ten-thousandth of an inch. Maudslay was also a fine teacher, and his workshop became a nursery for a group of talented young engineers.

machine tool – the screw-cutting lathe, the ancestor of all modern metalworking lathes, devised by British engineer Henry Maudslay. The screw-cutting lathe helped to pioneer the manufacture of highly accurate screw threads. Built on two parallel triangular bars, it incorporated a revolutionary new feature, the so-called 'leadscrew' between the bars that propelled the slide rest along. The slide rest carried a tool-holder for the knife that cut the thread, plus a micrometer dial to control the depth of cut precisely. With every turn of the 'workpiece' (the raw material that was to be made into a screw), the slide rest moved along. As it did so, the distance between the knife and the workpiece steadily decreased, shaving off a little more metal and cutting a deeper thread.

Maudslay's machines
An engraving taken from the Review of the Industrial Exhibition *of 1834 (above) showing various machines designed by Henry Maudslay. The screw-cutting lathe is highlighted in pink.*

THE FIRST ASSEMBLY LINE

The Block Mills built by the British Admiralty at Portsmouth Naval Dockyard were the first continuous assembly line in history. This factory was designed to rationalise the manufacture of wooden rigging blocks, some 130,000 of which were required every year, in varying sizes, by the Royal Navy's sailing ships. With the help of Henry Maudslay and others, from 1801 onwards Marc Isambard Brunel installed 45 purpose-built machines that produced blocks of superb quality. The savings in manpower were huge – 10 unskilled mill workers could turn out as many blocks in the same time as 110 master carpenters.

Maudslay's lathe not only helped to bring clockmaking into the industrial era, it also became emblematic of the machine-tool age.

Henry Maudslay trained a generation of British engineers whose names are synonymous with the growth of mass-production and machine tools, including Joseph Clement, Richard Roberts, James Nasmyth and Joseph Whitworth. Their inventiveness and skill bore fruit in the form of all kinds of lathes, hydraulic presses, shaping machines, machines for die-stamping and pressing, saws, planing machines, boring machines, grinders, milling machines, drill, corers, riveting machines, punching machines, mortising machines and more. By this time, machine tools were being

Fruitful collaboration
In 1803–5, Henry Maudslay and Marc Isambard Brunel built a machine to mechanise the manufacture of ships' blocks – the pulleys for the ropes that manipulate the sails. The machine (above) handled a job that previously involved 20 separate skilled operations.

powered by steam engines via networks of aerial pulleys and belt drives.

Now that machines were no longer dependent upon human muscle-power – which on average amounts to less than one-tenth of one horsepower, or equivalent to around 70 watts – their phenomenal work rate and ease of control enabled them to produce other machines. At the 1851 Great Exhibition, held in the Crystal Palace in London's Hyde Park, the huge range of machine tools proudly emblazoned with the legend 'Made in England' were the envy of the world.

America muscles in

Across the Atlantic, the United States was fast catching up. The country industrialised rapidly and a shortage of skilled workers prompted the development of a home-grown machine-tool industry. The first sector to embrace machine tools was the armaments industry. During the Civil War of 1861–5, Christopher Miner Spencer developed a fully automatic turret lathe that could mass-produce a new model of repeating rifle he had designed. Some 300,000 Spencer rifles were supplied to the Union Army. Other notable American inventions of the time included the Brown & Sharpe universal milling machine (1862) and William Gleason's bevel gear planer (1874).

National showcase
The stand of the engineering firm founded by Sir Joseph Whitworth (1803–87) at the Great Exhibition of 1851 (above). Ten years earlier, Whitworth had proposed a standard for the contour of screw threads (55°), which still bears his name.

Age of dynamism

Around the same time, the Swiss-German engineer Nicolas Junker set up a factory in the town of Moutier. There, in 1890, he perfected an automatic turning machine with a sliding headstock – a device that enabled the machine to simultaneously rotate the workpiece and move it along the spindle. Junker's aim was to supply high-precision miniature brass screws and pinions for the local watchmaking industry. Up until then, these parts were made laboriously, one by one, by hand-lathe operators.

Germany was another up-and-coming industrial nation in the 19th century. By 1895 it had developed its own thriving machine-tool

Heavy industry
Workmen constructing turbines in the machine hall at the AEG works in Berlin in 1900.

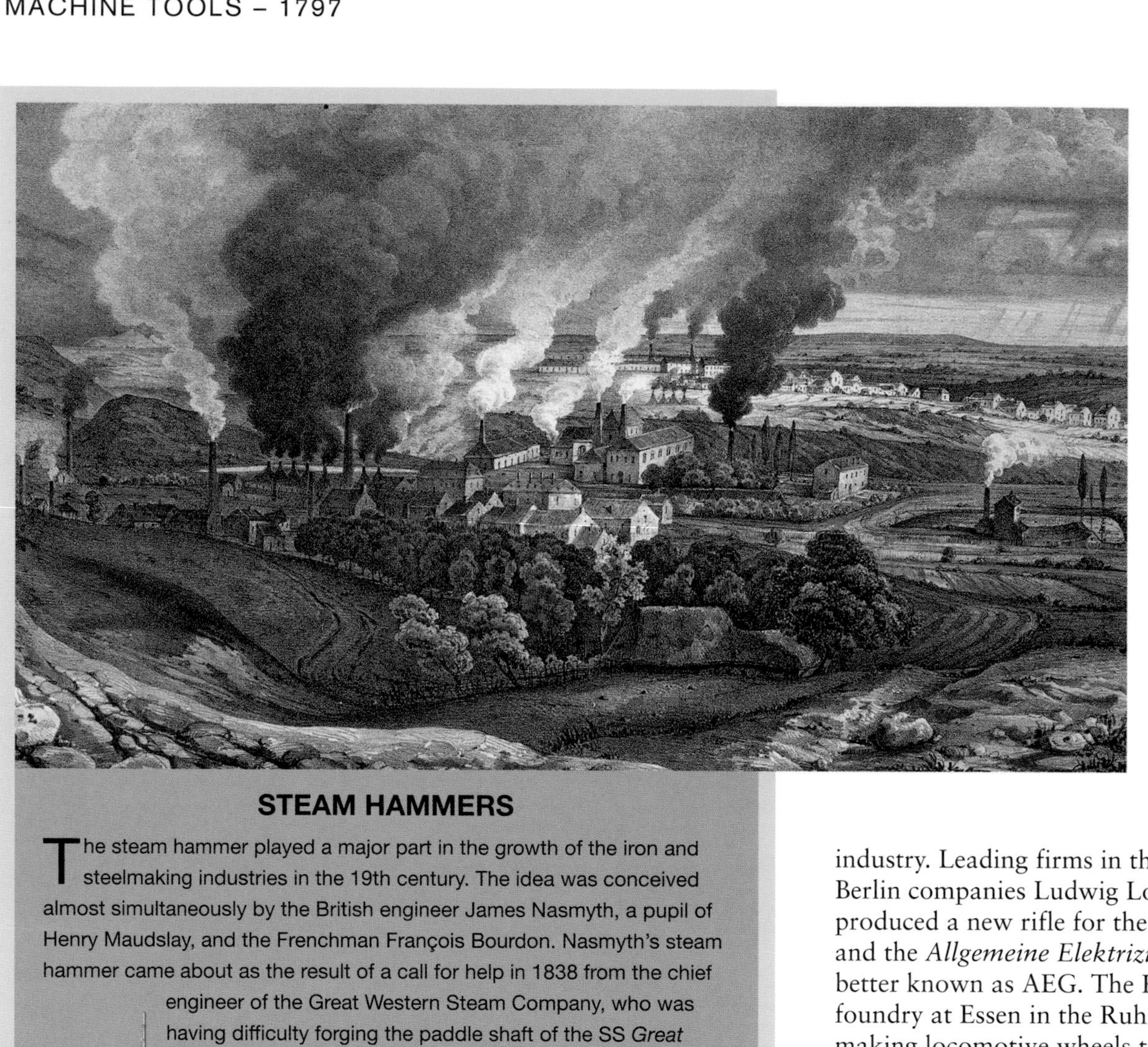

French steel town
An engraving of 1830 shows the iron and steel works at Le Creusot in eastern France. The town's metallurgical industry boomed with the establishment of the Schneider foundry there in 1836.France's first locomotive was built there in 1838 and its first steamship a year later. It was here that François Bourdon devised a method for manufacturing large metal components that could not be made by hand.

industry. Leading firms in the field included the Berlin companies Ludwig Loewe & Co (which produced a new rifle for the Prussian Army) and the *Allgemeine Elektrizitäts-Gesellschaft*, better known as AEG. The Krupp steel foundry at Essen in the Ruhr developed from making locomotive wheels to manufacturing armaments on a massive scale. Krupp invested heavily in new technology, including in 1861 a steam hammer nicknamed 'Fritz' that could deliver a 50-tonne blow.

By the end of the 19th century, the power, precision and output of machine tools far outstripped the capacity of manual labour. All sectors of manufacturing now relied on them. The new technologies of the steam locomotive and ocean-going steamships brought a further step-change in the power and size of machine tools in order to produce the huge engines and massive components they required.

Towards full automation

Mechanisation increased industrial productivity, drove down the price of finished goods and maximised employers' profits. But its negative social impact could not be overlooked. In early 19th-century England, mechanisation had been the trigger for a wave of destruction of automated looms and stocking frames by weavers, the 'Luddites', who believed that the machines threatened their livelihood. And as mechanisation grew in scale, many workers found themselves surplus to requirements.

In the early 20th century, manufacturing industry was revolutionised for a second time

STEAM HAMMERS

The steam hammer played a major part in the growth of the iron and steelmaking industries in the 19th century. The idea was conceived almost simultaneously by the British engineer James Nasmyth, a pupil of Henry Maudslay, and the Frenchman François Bourdon. Nasmyth's steam hammer came about as the result of a call for help in 1838 from the chief engineer of the Great Western Steam Company, who was having difficulty forging the paddle shaft of the SS *Great Britain* because the tilt of the hammer he was using was not sufficiently versatile. Nasmyth sketched out an idea in his 'Scheme Book' and subsequently went on to produce his steam hammer, which he patented in 1842, at his Patricroft foundry in Manchester. One of the key features of the new hammer was that the operator had complete control over the force of each blow. Naysmith enjoyed demonstrating how it could, at one extreme, break an egg placed in a wine glass without damaging the glass, and at the other extreme produce a blow with enough force to shake a building. Naysmith went on to develop steam pile-drivers. Bourdon is thought to have used Naysmith's design to build his original steam hammers at the Schneider ironworks at Le Creusot in France.

Mighty beast
The Creusot steam hammer operated from 1877 to 1930, and was for a long time the most powerful in the world. It stands in the centre of Le Creusot as a monument to the town's iron and steel heritage.

by the advent of the electric motor. Writing in 1909, a commentator speculated that 'electric motors powering each and every machine tool will surely soon be the order of the day'. This new technology made machine tools even faster and more accurate. Around 1910, multiple drill presses were introduced in the USA. Henry Ford installed them on the assembly line at his factory at Highland Park, Michigan, to drill gearbox casings for the Model-T car.

The development of new types of steel improved cutting speeds, while the introduction of more durable machine tools made from tungsten carbide further increased efficiency. Large-scale automation of manufacturing began in the interwar period, and was given new impetus in the 1970s as the IT revolution took hold. In certain processes, the advent of computer-assisted manufacturing would lead to human input being reduced to supervisory and maintenance roles, and the ongoing trend towards greater automation has continued unabated in the 21st century.

Nightmare vision
In 1927 German film director Fritz Lang's science-fiction dystopia Metropolis *(below) portrayed a futuristic world in which humans were enslaved by machines.*

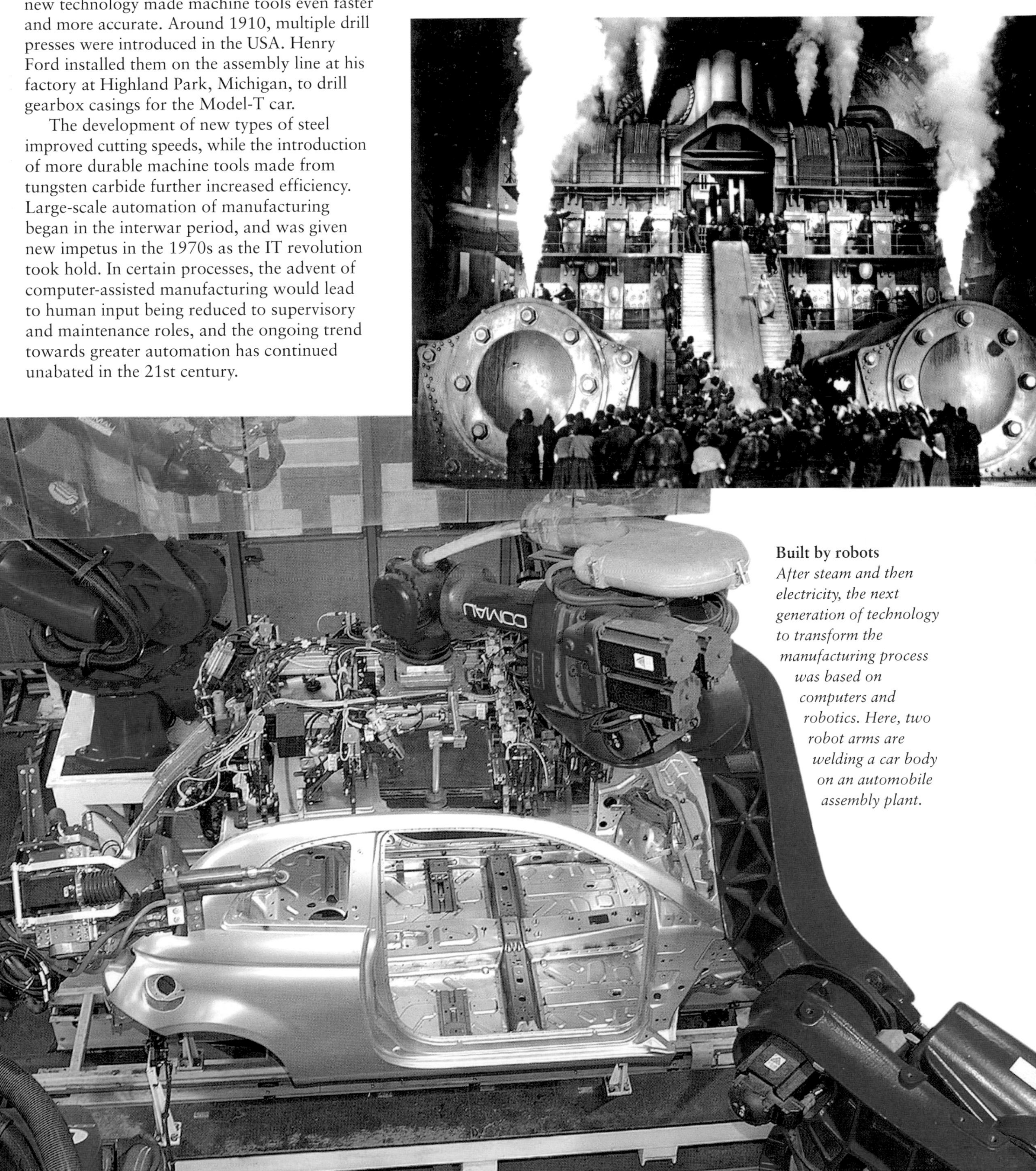

Built by robots
After steam and then electricity, the next generation of technology to transform the manufacturing process was based on computers and robotics. Here, two robot arms are welding a car body on an automobile assembly plant.

The heights of virtuosity

The heyday of classical music, spanning the half century or so between the Baroque and the Romantic eras, may be said to begin with the death of J S Bach in 1750 and end with that of Haydn in 1809. This brief but fecund period left an indelible mark on the history of music.

Court composer *A model of the main theatre at Esterháza Palace in Hungary from c1770 (above right), the period when Joseph Haydn was employed there as Kapellmeister to the Esterházy princes. He was required to organise up to 100 musical soirées a year for their pleasure.*

The year 1750 saw the death of the great and prolific German composer Johann Sebastian Bach. The work he was writing at the time, *The Art of Fugue*, represents the high-water mark of the Baroque – a musical style characterised by the use of counterpoint. Yet even while such composition was reaching a pinnacle of sophistication in Bach's fugues, a revolution was underway that would change the face of music. A growing popular taste for a lighter, more immediately entertaining style of music was becoming apparent. Virtuosity was becoming all the rage, as the appeal of melody began to win out over intellectual complexity.

Ensemble playing *Haydn conducting a string quartet in Vienna, from a 19th-century lithograph. He is credited with creating this form of ensemble.*

Composers abandoned abstract, polyphonic structures in favour of a more down-to-earth approach, marked above all by a clear distinction between the lead instrument and the accompanists. Symmetry and balance were the order of the day. Each piece developed from a limited number of concise motifs, whose repetitions and variations were governed by the largely set structure of the sonata form. The era of classical music had dawned.

This stress on simplicity and clarity means that works of this period tend to lodge in the mind of the listener. Certain tunes by late 18th-century composers like Haydn and Mozart are known the length and breadth of Europe. One of Haydn's melodies became the German national anthem.

The rise of instrumental music

The popularity of instrumental music brought some hitherto neglected genres to the fore – the concerto, the symphony and chamber music.

TONALITY

One defining feature of classical music is its tonality. The musical system established hierarchical pitch relationships between notes by setting a distinct tonal centre or 'key' to the composition. The foundation stone of tonality is harmony, the combining of three simultaneous notes into a chord, with the root note being referred to as the 'tonic'. So, in a piece written in D, the music can modulate into other keys but will always return to the underlying tonality (or 'home key') of D.

THE GENESIS OF OPERA

The earliest surviving opera dates from around 1600. The first acknowledged masterpieces of the form are the work of Venetian composer Claudio Monteverdi in the early 17th century. The vocal style in Italian opera was known as *bel canto* ('beautiful singing'), hallmarked by agile, virtuosic and florid delivery. Opera gained a firm foothold in Venice and Naples, then spread to London, Vienna and Paris. Its main exponents in the classical period were Gluck and Mozart. Beethoven composed only one, *Fidelio*, in 1814.

Stage and screen *A scene from Mozart's final opera* The Magic Flute *(1791), as filmed by the Swedish cinema and opera director Ingmar Bergman (right).*

The huge output of symphonic works in the second half of the 18th century – Haydn alone wrote 104 – set the pattern for the modern classical orchestra built around the nucleus of a string section comprising violins (divided into first and second violins), violas, cellos and double basses. The wind section was made up of woodwind (flutes, oboes, clarinets and bassoons) and brass (horns, trumpets, and – in the case of Beethoven – trombones). Finally came the percussion section. Orchestras of this period were generally conducted from the piano or lead violin by a Kapellmeister.

Chamber music, as the name suggests, developed as a form for more intimate settings, such as musical soirées or salons. The term came to denote instrumental music for small ensembles of players, ranging from soloists to octets. Works were popularised by music publishers selling printed sheet-music scores.

New genres also began to emerge, such as the piano sonata and the string quartet. The pianoforte, which first appeared in the early 18th century, gradually supplanted the harpsichord. The first musician to write specifically for the piano was the Anglo-Italian composer Muzio Clementi. In the hands of virtuoso performers and composers such as Mozart and Beethoven it became a firm favourite with concert-goers. Meanwhile the string quartet, comprising two violins, a viola and a cello, was characterised by a subtle balance and interplay between the four parts, and the string ensemble quickly caught on across the whole of Europe.

Increasingly, Italian musicians such as Antonio Salieri and Germans like Christoph Willibald Gluck began to gravitate to Vienna, home of the Habsburg court. It was from there that the three greatest figures in classical music – Haydn, Mozart and Beethoven – were all presently to emerge.

Haydn and Mozart

The long life of Joseph Haydn (1732–1809) represents a vital link in the story of music. Born while J S Bach and Vivaldi were still alive, Haydn's own death coincided with the birth of the first Romantic composers. Haydn was hugely popular in his own day and was fêted throughout Europe. He spent the final years of his life in Vienna, where he wrote his late masterpieces and acted as a mentor to young composers, among them Ludwig van Beethoven. No composer before Haydn exerted so great an influence on his contemporaries. His many symphonies and his chamber music were model examples of their respective forms. It was Haydn who established the sonata sequence of four movements with contrasting tempos – *allegro, andante, minuet* and *allegro* or *presto* – and who laid the foundations of the classical mode.

The meteoric career of Wolfgang Amadeus Mozart (1756–91) held his contemporaries spellbound and continues to fascinate. The son of Leopold Mozart, himself a composer, the young Wolfgang was exhibited as a prodigy by his father around the courts of Europe. He played the piano and violin flawlessly and was writing his own compositions from the age of

Boy wonder *A bronze statuette of the young Mozart by the sculptor Louis Ernest Barrias in 1887.*

five. Mozart was not a musical innovator; his genius was to take existing forms to the peak of their possibilities. His Italian operas *Don Giovanni*, *The Marriage of Figaro* and *Così fan Tutte* took the genre to new heights of perfection, while his final operatic work *The Magic Flute* broke new ground with a libretto in German. His premature death in poverty, while composing his unfinished *Requiem*, helped to foster his posthumous legend.

Eve of Romanticism

The last of the classical triumvirate is Ludwig van Beethoven (1770–1827), who straddled the 18th and 19th

THE SONATA

The term 'sonata' denotes an instrumental piece, played by a soloist or a small ensemble. The genre was popularised by Haydn, Mozart and Beethoven. The sonata has also given its name to the 'sonata form', a method of musical composition widely used since the early Classical period. This consists of an exposition (an enunciation of two contrasting themes), a development (transformation and variation of the themes) and recapitulation (return to the main theme and affirmation of the underlying tonality). The sonata form, which comprises the basic framework of most classical works, relies on the listener's ability to recall the themes and pick up on their repetition and variation.

Opera star
Giuseppe de Begnis, a renowned 19th-century Italian bass, in costume (above) for the title role in Mozart's Marriage of Figaro, *performed at the King's Theatre in London in 1823.*

Maestro at work
Beethoven at the piano, from an 1890 engraving by Carl Schloesser. It was said that Mozart could compose an entire work in his head while playing billiards. By contrast, even before his deafness set in, Beethoven reputedly agonised over every note.

Symphony in D
A handwritten score of Beethoven's Ninth Symphony, the 'Choral', in D Minor. The work was written between 1822 and 1824 and premiered at the Kärtnertortheater, Vienna, on 7 May, 1824.

centuries. Like Mozart, his talents as a virtuoso pianist and composer became clear at an early age. He gradually lost his hearing around the age of 30 and gave up public performance a decade later, but he continued to compose and conduct almost until his death.

After settling in Vienna in 1792, Beethoven performed and published to great acclaim, both in the Austrian capital and beyond – in Prague, Dresden, Leipzig, Berlin. His work had a huge impact on his contemporaries. He brought orchestral composition into the modern age and set new standards for piano composition and playing. From his Third Symphony, the 'Eroica' (1803) onwards, when he was struggling with growing deafness, a new aesthetic began to emerge which broke the bounds of Classicism. The first stirrings of Romanticism can be heard in the emotional depth and exuberance of the 'Eroica' and in his later chamber pieces. His final string quartets display a profound spirituality, while his piano sonatas paved the way for later virtuosos such as Schubert, Chopin and Liszt.

THE CONCERTO

The concerto is characterised by interplay between the orchestra and one or more solo instruments. It originated in Italy in the early 18th century and its emphasis on virtuosity soon gained it a widespread following. The most common forms are the violin or piano concerto; Mozart's piano concertos represent a pinnacle of perfection though he also wrote for the bassoon, clarinet and French horn, while Haydn composed cello concertos.

Piano virtuoso
The concert pianist Alfred Brendel (left) was born in Czechoslovakia in 1931, but lived in England from the early 1970s. His wide repertoire ranges from Bach to Schoenberg. Brendel was knighted in 1989 for his services to music. He retired from concert playing in 2008.

THE JACQUARD LOOM –1800

Weaving on an industrial scale

The loom and weaving process introduced by Joseph Marie Jacquard changed textile manufacturing from a cottage industry into a full-blown industrial enterprise. His invention not only made weaving less labour-intensive, it also enabled manufacturers to produce the most complex patterns and foreshadowed the rise of the computer more than 150 years later.

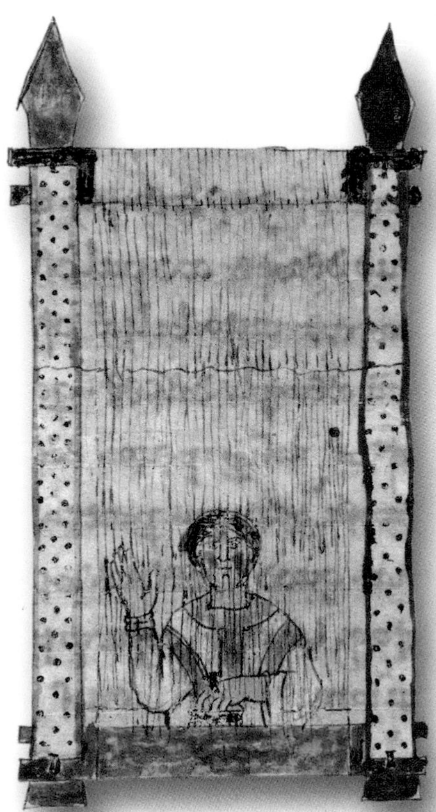

Manuscript miniature *An illumination of a weaving loom from a 9th-century work* De Rerum naturis *('On the Nature of Things') by the German monk Rabanus Maurus.*

In the late 18th century, the vast majority of weaving was done by hand. Work was contracted by central agents to artisans, who wove cloth from raw silk thread, cotton yarn or wool in their homes on hand-looms. Weaving was often a part-time occupation to supplement the family's income from farming. As a result, textile manufacture was a diffuse business, with agents collecting pieces of cloth from scores of home weavers and selling it on.

A laborious operation

Clothmaking had seen a number of advances. In 1605 a Milanese weaver named Claude Dangon, working in the French silk-weaving centre of Lyons, streamlined the drawloom by adding a drawing-fork (or lever) and trebling the number of vertical and horizontal draw-cords (or 'leashes') that raised or lowered alternate warp threads of the fabric. His lever drawloom allowed greater complexity in the woven pattern and remained in use for complex damasks and heavy brocaded cloth until around 1800. But to operate this type of loom the weaver required assistants to lift the warp and pass the shuttle containing the weft through the gap (the 'shed').

In 1733 the British inventor John Kay introduced a piece of apparatus that greatly simplified this process: the flying shuttle. This device allowed the weaver to 'throw' the shuttle through the shed across the whole width of the cloth, and then quickly pull it back using a fine cord attached to the shuttle. Kay's invention was a major step forward, speeding up the process of weaving and making it less labour-intensive. It also reduced the number of imperfections in the finished fabric and made it possible to weave wider bolts of cloth, which hitherto had been restricted to the span of the weaver's arms.

Despite these improvements, textile manufacture remained a laborious business, particularly for the draw-cord operators who, being mainly children, were known as 'drawboys'. The flying shuttle downgraded

Village handicraft *A Mexican weaver makes a rug on a simple hand loom. In countries where industrialisation was late to develop, weaving techniques have remained largely unchanged from those used in the Middle Ages.*

INTERDEPENDENT TECHNOLOGIES

John Kay's flying shuttle, invented in 1733, accelerated the weaving process, allowed wider pieces of cloth to be woven and halved the number of people required to operate a loom from two to one. Even so, manufacturers were slow to embrace this important innovation in the mechanisation of clothmaking. Spinning and weaving go hand-in-hand and the spinners, who still worked by hand, were struggling to supply sufficient raw yarn. By increasing the weaver's workrate, the flying shuttle only made this problem worse and caused the price of yarn to skyrocket. The situation was not resolved until the advent of the spinning machine, the prototype of which was unveiled by Birmingham manufacturers Lewis Paul and John Wyatt in 1738. Then, in 1769, Richard Arkwright patented the water frame – a spinning machine driven by water power. The final piece in the jigsaw fell into place ten years later, with Samuel Crompton's 'Spinning Mule', which could spin thread simultaneously onto a thousand spindles.

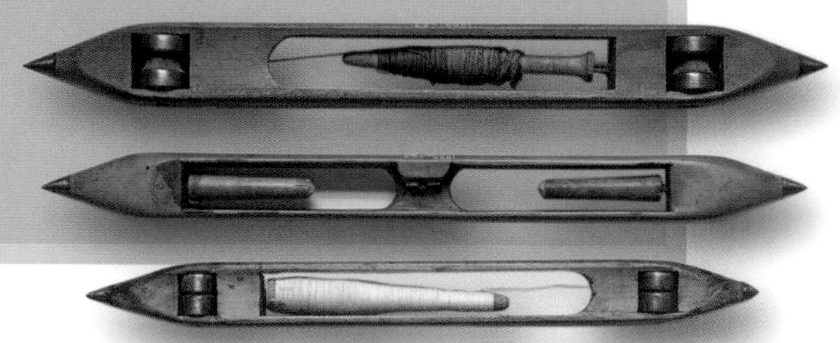

their lowly status further, while a huge increase in the demand for textiles in the 18th century exposed a lack of available manpower.

Also in Britain, in 1785 Edmund Cartwright patented the first power loom – essentially a steam-powered, mechanically operated version of a regular loom – and set up a factory in Doncaster for manufacturing cloth. So by the time French engineer Joseph Marie Jacquard of Lyons introduced his programmable loom in 1800, textile manufacture was already evolving from a cottage industry into an industrial-scale enterprise. What Jacquard's invention did was to sweep away the antiquated paraphernalia of leashes and treadles that required several people to operate. His ingeniously simple mechanism enabled a single operator to turn out the most complex patterned cloth.

Pioneers of automation

Jacquard was by no means the first person to conceive of programming weaving machinery. In around 1725 another weaver from Lyons, Basile Bouchon, had devised a semi-automatic loom in which the lifting and lowering of warp threads was controlled by a continuous loop of perforated paper tape. The warp threads passed through the eyes of horizontal hooked weaving needles arranged to slide in a box. These were either raised or lowered depending on whether the tape had a hole in it at that point. In this way, the loom operator could repeatedly weave the same pattern into the cloth at regular intervals. The perforated paper-roll mechanism was similar to that used in the automatic 'player-pianos' that became popular in the late 19th century.

For all its ingenuity, Bouchon's loom only used a single row of hooks. In 1728, Bouchon's assistant Jean-Baptiste Falcon replaced the fragile paper tape with a chain of interlinked punched cards, which could be swapped around to alter the pattern. This improved machine could move several rows of hooks at the same time and manipulate up to 400 warp threads.

Nevertheless, it still required an operator to feed the cards in manually. In 1744, Jacques de Vaucanson, official inspector of French silk manufacture and a renowned constructor of automata, modified Bouchon's system with a key innovation. In his design, a perforated

Falcon's semi-automated loom
On this ingenious machine, built in 1728, each row of vertical or horizontal holes corresponds to a line of stitching in the pattern on the cloth, in either the warp or the weft thread.

Intricate mechanism
A close-up view of the perforated drum mechanism on Vaucanson's 1744 automatic loom. Vaucanson applied the same techniques he had used in building automata to improving the loom.

drum regulated the movements of the warp threads, thereby making the loom fully automatic. A crank attached to a power source (either a horse wheel or waterwheel) activated the complicated series of cams, ratchets, connecting rods and gears. 'With this machine,' boasted Vaucanson, 'a horse, cow or donkey can make even more beautiful and perfect fabrics than the most highly skilled silk weaver'.

Unfortunately, Vaucanson's system turned out to be far too complex and only succeeded in repeating the same pattern over and over again. It was also prohibitively expensive to adapt it to fit existing looms. As a result, Vaucanson's loom never got beyond the prototype stage, but it did pave the way for Jacquard.

One-man operation

Jacquard was a shrewd engineer, who incorporated in his loom many of the innovations of previous inventors. In 1800 the National Academy of Arts and Crafts, of which he was a member, engaged him to review their collection of earlier weaving technologies. In the process, he came across the dismantled parts of Vaucanson's original loom and decided to rebuild it. He also revived Bouchon and Falcon's principle of punched cards, which he linked to the automatic drive of Vaucanson's perforated drum. This was the genesis of his famous loom, which was three times the size of the machines built by his predecessors. From the perforated cards, the Jacquard loom 'reads' the pre-programmed design that is to be woven into the fabric. The threads are then raised and lowered automatically. This effectively put an end to the onerous task of operating the draw-cords. With every stoke of the loom, a complex system of pulleys and counterweights flattens the punched card against a pipe-shaped drum (a refinement of Vaucanson's cylinder); by pressing on a treadle, the operator moves the card on one notch. Hooked weaving needles either raise or lower the harness carrying the warp thread, according to whether or not they are in contact with the holes in the card, and the process carries on incessantly. The weaver is the sole person required to operate the loom.

In 1805 Jacquard unveiled a loom with no fewer than 1,200 hook heads, which was capable of weaving the most complicated patterns. Geometrical designs or flower patterns, for men's waistcoats and women's dresses alike, were very much in vogue at the time. Right from the outset, textile manufacturers eagerly embraced the new technology. The faster output and reduced workforce that it entailed meant that the

Historic fabric
A detail from a silk fabric manufactured in 1809 by the Dutillieu mills in Lyons (founded 1761), from the collection at the city's Museum of Textiles and Applied Arts (right).

Labour-saving device
A 19th-century engraving showing an operator at an 1801-model Jacquard loom. Jacquard was deeply troubled by the social impact of his invention, which brought in its wake a wave of unemployment in the textile industry.

Automated process
A pile of punched cards used in Jacquard looms. Each row of holes corresponded to one row of the woven design.

A RELUCTANT CELEBRITY

Joseph Marie Jacquard was born into a family of weavers in Lyons on 7 July, 1752. Little is known about his early life, except that his brother-in-law, who was a bookseller and printer, got him an apprenticeship as a bookbinder. He led an entirely unremarkable existence until his late forties. Then, in 1800, he submitted a patent for a labour-saving loom that he had designed. On 12 April, 1805, he was catapulted to fame when Napoleon I and Empress Josephine viewed his machine while visiting Lyons. In August of that year, the Lyons Academy bestowed on Jacquard the first of many honours and prizes. In 1806, in return for making his loom public property, the city granted him a pension for life and a royalty on each machine sold – his workshop turned out 57 looms in 1808. Jacquard himself gradually lost interest in his invention and spent less and less time at his workshop. Even so, the city fathers feared that he might take his skills to another textile centre, and continued to shower him with gifts.

By 1812 there were around 11,000 Jacquard looms in use in France. Increasing mechanisation prompted a major revolt by Lyons weavers in 1831. Jacquard died at his country house in Ouillins, which he briefly represented as a town councillor, on 7 August, 1834.

Reviled inventor
An engraving from Martyrs of Science *by G Tissandier, made in 1882 (below). On more than one occasion, Jacquard was attacked by Lyons weavers, angered by the automated loom that robbed them of their livelihoods.*

WEAVERS' REVOLTS

The changes in working practices brought on by mechanisation often led to serious industrial and social unrest. In Britain, the most famous uprising in the textile industry was the Luddite revolt of 1811–12. Taking their name from the fictitious Ned Ludd, a working-class hero, hand weavers set about wrecking the new mechanised looms. The revolt spread quickly through Nottinghamshire, Yorkshire and Lancashire, before being quelled by the army. The ringleaders were hanged. The term 'Luddite' is still used scathingly of anyone who opposes technological change. In late 1831, the French city of Lyons saw serious unrest, prompted by a drastic fall in the price of woven silk after the introduction of the Jacquard loom. Weavers seized the town hall and proclaimed a rebel government. In Germany, a major revolt by Silesian weavers in 1844, against a background of rising unemployment and falling living conditions, marked a key moment in that country's labour history. Karl Marx saw in it the birth of an organised workers' movement. Later it was the subject of a highly influential play, *The Weavers*, by the dramatist Gerhart Hauptmann.

productivity of mills soared. On a Jacquard loom, a single weaver could turn out up to 50 metres of finished cloth a day. The merits of the loom were quickly recognised in other countries. By 1833 there were about 100,000 power looms in use in Britain that were heavily influenced by Jacquard's invention.

Industrial unrest

The increased productivity of the new looms came at a price paid by the workforce. It was estimated that every Jacquard loom made five drawboys redundant. The loss of jobs provoked violent hostility among textile workers. Jacquard's home town had seen the public burning of one of his looms as early as 1806. Jacquard himself narrowly escaped being thrown in the River Rhône by the mob, and had to be rescued by the local militia. But the march of progress was unstoppable.

Jacquard's loom subsequently underwent several modifications. Early examples were somewhat erratic and prone to breakdowns.

Factory inspection
Visitors to a textile mill in the 19th century examine an automated loom (below). In the weaving shed, rows of identical machines like this one would have been powered by a central steam engine via a system of belts and pulleys.

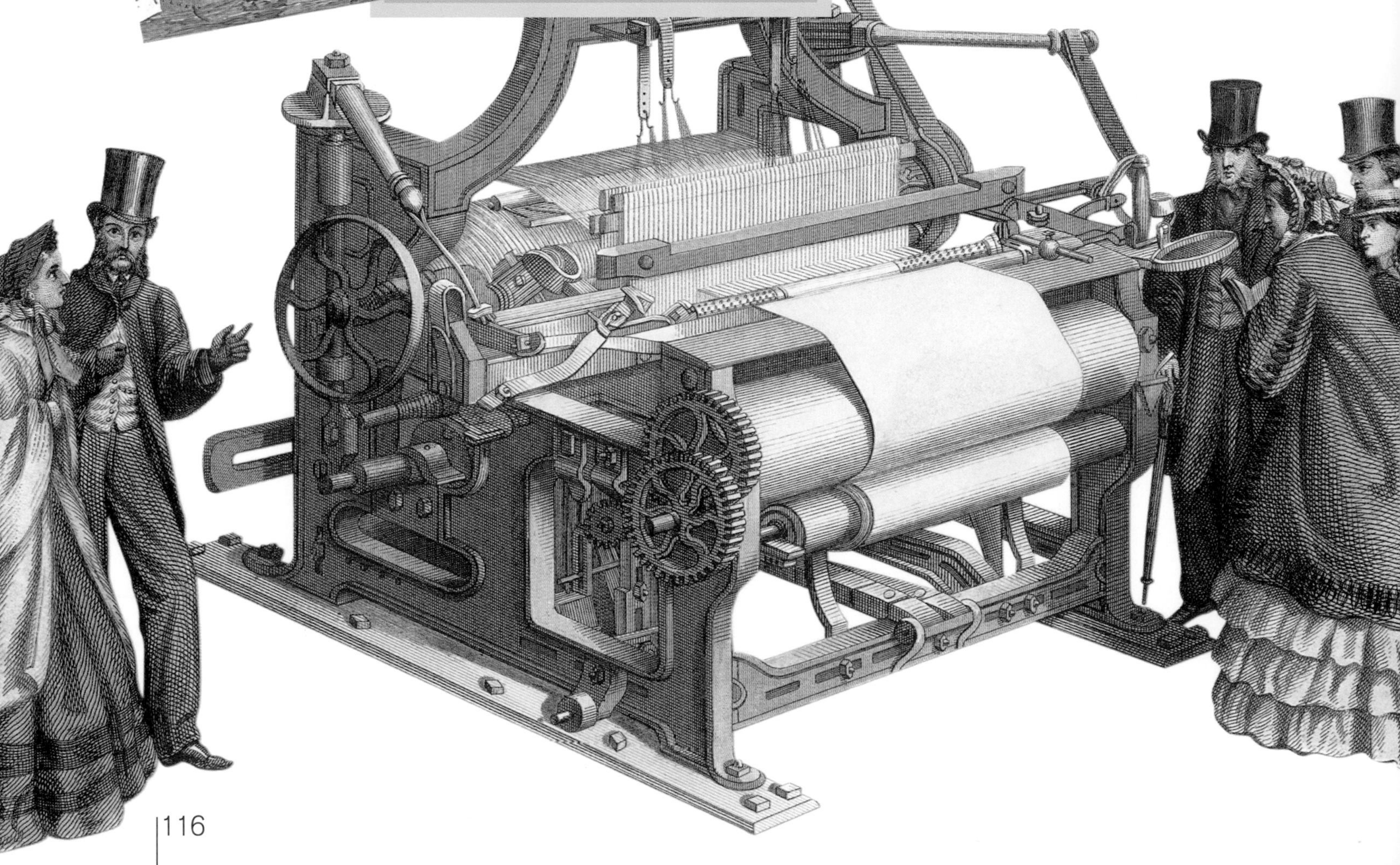

Every imperfection in finished cloth meant a reduction in wages for the weaver. Another engineer from Lyons, Jean-Antoine Breton, made some decisive alterations to Jacquard's machine, most importantly the addition of an angled flange that pressed the punched card firmly down onto the drum. It was this modified machine, still known as the Jacquard loom, that was widely adopted after 1816.

The second half of the 20th century saw the introduction of shuttleless electric looms, which have made weaving even faster. According to the particular model, the weft thread on electric looms is thrown either by a blade or a jet of compressed air or water. Even so, the Jacquard loom did not disappear. It is still in use today in the manufacture of complex patterned fabrics such as brocades and damasks. Once a byword for mass production, it has now become associated with artisan weaving.

Aside from its role in textile manufacture, the Jacquard loom occupies an important place in the history of technology. Its punched-card operating system made it the grandfather of the modern digital computer.

Modern weaving
The Tassinari and Chatel textile mill in Lyons (top), suppliers of silk to all of France's national museums. Electric looms weaving madras (inset), a lightweight fabric combining silk warp threads with a cotton weft.

THE LOOM AND THE ANALYTICAL DIFFERENCE ENGINE

Cambridge professor of mathematics Charles Babbage (1792–1871) is widely recognised as the pioneer of computers. From 1833, he devoted much time and energy to the invention of an 'analytical difference engine' – a programmable computing machine. The punched-card technology that Babbage employed in his machine was very similar to the operating system of early computers, which began to appear in the 1940s. These cards effectively contain binary commands, according to whether perforations are absent or present – the same principle used in the Jacquard loom. Babbage freely acknowledged his debt, writing 'there is an almost complete analogy between the analytical engine and the [Jacquard loom] process'. His collaborator Ada Lovelace, daughter of Lord Byron and an eminent mathematician in her own right, said it even more plainly: 'We may say most aptly that the analytical engine weaves algebraical patterns just as the Jacquard loom weaves flowers and leaves'. Babbage's machine was designed to run on steam, but sadly he never managed to iron out flaws in the mechanism. He died in poverty.

The mechanised loom

Inventions rarely appear from nowhere, fully fledged. Almost invariably, they build on a sequence of earlier developments. The Jacquard loom is a classic example, being the fruit of a process of technological evolution that had been going on for centuries. In its turn, it helped give rise to new technologies such as the computer.

An ancient spindle for spinning wool.

SPINNING
FROM THE WHORL TO THE ROTOR

In *c*7000 BC, spinning was done with a distaff and a wooden spindle mounted on a heavy disc of stone or clay – the whorl. The spinning wheel, invented in Asia in the 5th century AD, was brought to the West by the Arabs in the 13th century. There, it evolved into the double-drive spinning wheel, powered by a hand crank or a treadle. In the 18th century, the spinning jenny (1764), water frame (1768) and mule jenny (1779) brought spinning into the industrial age. The ring-spinning machine (invented in the US in 1828) could twist 6,000 metres of yarn onto multiple bobbins, which were automatically replaced once fully wound. Rotor spinning was invented in Czechoslovakia in the 1960s; in this process, a rotor revolving at up to 140,000 rpm catches strands of fibre and twists them into perfect yarn.

Hand looms are still used for handicraft weaving of cotton and other fabrics.

TEXTILES
FROM NATURAL TO SYNTHETIC FIBRES

Arable farming and animal husbandry produced natural fibres such as linen, cotton, hemp, wool and silk. The first synthetic fibres only appeared in the late 19th century. In the 20th century, the petrochemical industry developed a range of new synthetics, including polyamides (nylon), polyesters (Terylene, Dacron) and acrylic fibres. Insulating, non-iron and hard-wearing, they are often combined with natural fibres to make easy-care fabrics.

Skeins of dyed wool ready for weaving.

A close-up of a punched card, as used in the Jacquard loom.

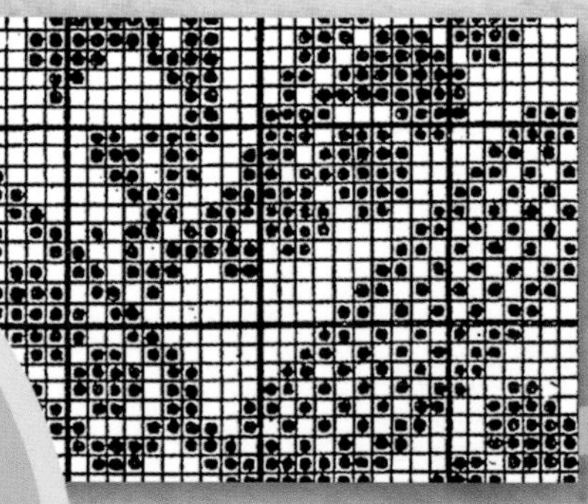

AUTOMATION
FROM THE JACQUARD LOOM TO THE COMPUTER

Before Jacquard's programmable loom, weaving patterned cloth was a labour-intensive business, requiring drawboys to repeatedly lift or lower the harness carrying the warp thread. In an automated loom, the number of threads was determined by the pattern, while the weave was controlled by perforated cards prepared by punching holes in a piece of thick square-ruled paper. The punched-out squares represented the pattern and the unpunched squares plain fabric. Vertical lines of holes were for the warp threads, horizontal for weft. When the loom's control-rod met a hole, it passed through to make a stitch. Computer technology took a cue from this card-punch method of programming a machine.

Modern computers like the IBM JUBL supercomputer (right) can perform 37.3 billion calculations per second.

THE SHUTTLE

FROM THE SHUTTLE
TO THE AIR-JET

In the 17th century, wooden shuttles carrying spools of yarn came into use for passing the weft thread between the warp. John Kay's flying shuttle (1733) had four rollers to reduce friction. In automated looms, the shuttle lost its rollers and became larger. On Jacquard looms, the shuttle box could be programmed to make the colour change of the weft thread automatically. The shuttle was supplanted altogether in looms in the 20th century, which use jets of water or air to take the yarn through the shed.

A box of flying shuttles.

WEAVING

FROM HAND WEAVING
TO COMPUTER-GUIDED LOOMS

Weaving derived from basketmaking. The earliest known portrayal of a loom is on an Ancient Egyptian pot dating from 4400 BC. The pedal-operated loom was invented in China in *c*2000 BC. The drawloom, originally from the Middle East, enabled more complex patterns to be woven. Its operation was improved by Frenchman Claude Dangon in the lever drawloom of 1605. In 1725 Basile Bouchon introduced a perforated paper command system that partially automated the weaving of patterned cloth. In 1728 Jean-Baptiste Falcon replaced the paper strip with punched cards, and in 1744 Jacques de Vaucanson devised a perforated drum mechanism to raise and lower the warp. In 1800 Joseph Marie Jacquard combined all these innovations in his automated loom. The next step came with powered looms. Today, looms are computer-guided, requiring only one operator to supervise several machines.

The lever drawloom, in use from the early 17th century onwards.

Stone bas-relief of a weaver from the second millennium BC.

The Jacquard loom.

POWER SOURCES

FROM STEAM POWER
TO ELECTRIC POWER

Steam engines began powering looms with the invention of Edmund Cartwright's power loom in 1785. Power was transmitted from a single central engine via shafts suspended from the ceiling and linked by belts to individual looms. At the start of the 20th century, steam engines were superseded by large electric motors. By the late 1940s, in most mills each loom had its own small electric motor as an independent power source.

Electric looms in a modern textile factory.

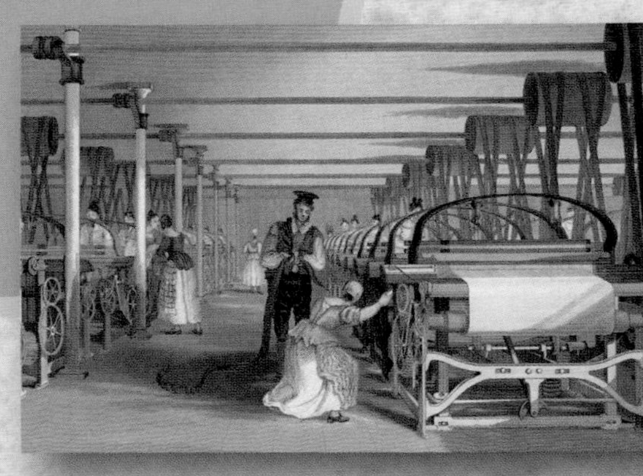

Steam-powered looms in a 19th-century textile mill.

INFRARED LIGHT – 1800

Studying the spectrum

Prompted by his observations of the Sun, the German-born astronomer William Herschel devised an experiment that was to have far-reaching implications. He discovered an invisible form of light later identified as infrared radiation. Many practical applications have followed from this discovery, from remote-control to night-vision devices.

Refracting light
A glass prism used by Herschel in his experiments on thermal radiation in the solar spectrum (left).

By appointment to His Majesty
Following his discovery of the planet Uranus in 1781, Herschel (right) was knighted and made personal astronomer to George III. An annuity of £200 per year granted by the King enabled Herschel to pursue astronomy full-time. Seen behind him in this portrait is the giant 40ft (12m) telescope that he erected at his house in Slough.

In 1800 Sir William Herschel (1738–1822) began a series of experiments on light. Some decades earlier, the scientific community had become embroiled in a sterile theoretical debate concerning the nature of light: was it composed of a mass of tiny particles (or 'corpuscles', as proposed by Newton) or did light waves travel on an invisible ether (the view of Dutch physicist Christiaan Huygens)? Herschel's concern was far more practical in nature. He wanted to see whether he could pick up the light from stars in his telescope. Then in his sixties, Herschel was already an astronomer of great renown, famous for his 1781 discovery of Uranus – then the furthest known planet in the solar system – and many other celestial bodies.

Beyond red

In one series of observations, Herschel used his telescope during the day to gather data on the Sun. The blinding strength of the Sun's rays made solar observation a dangerous business. Using different coloured lenses, Herschel found that there was no direct correlation between the Sun's light intensity and the heat that it radiated. He noted in his journal: 'When using particular ones [coloured lenses] I could feel plenty of heat in spite of the small amount of light they let through; others, by contrast, let through plenty of light but almost no heat.'

He decided to measure precisely, with the aid of a prism, how much heat was given off by each colour of the spectrum. He split white light into its individual components – the colours of the rainbow – and concentrated their rays on thermometer bulbs. Straightaway he noticed that the temperature was highest at

RITTER'S 'CHEMICAL RAYS'

In 1800, following his discovery of infrared, Herschel began to investigate the other end of the spectrum, beyond violet, but he found no significant variation in the temperature. The very next year a young German physicist, Johann Wilhelm Ritter (1776–1810), discovered ultraviolet light while experimenting with silver iodide, a chemical that turns black in sunlight. Placing the iodide just beyond the violet end of the spectrum, he was surprised to see it quickly darken. This indicated the presence of a form of light, which Ritter called 'chemical rays'.

the red end of the spectrum, but he looked on with astonishment as it kept climbing beyond the red, where there was no apparent light. Herschel had discovered an invisible form of radiation: infrared. Speculating on the nature of this phenomenon, he came up with various incorrect hypotheses. Only some time later, in 1865, did the Scottish physicist James Clerk Maxwell demonstrate that infrared radiation is one of a number of electromagnetic waves that form the electromagnetic spectrum, comprising radio waves, visible light, X-rays, gamma-rays, ultraviolet and others.

Practical applications

Different wavelengths of infrared radiation are now used in a wide variety of applications, from remote-control devices to car key-fobs to heating elements in microwave ovens. Infrared technology is also used in night-vision equipment: the faintest source of heat emits infrared radiation, revealing the presence of a warm body, even in pitch-black darkness. Appropriately enough, modern infrared telescopes have also given astronomers a totally different view of the universe, revealing the existence of huge clouds of dust and gas within our solar system. These give off no visible light but, warmed by the stars, they still emit infrared radiation. In April 2009 the European Space Agency launched a new telescope to observe the universe in infrared. It was named 'Herschel' in homage to the astronomer who first identified the phenomenon.

INFRARED AND THE GREENHOUSE EFFECT

The Earth's surface reflects back part of the heat generated by the Sun's rays in the form of infrared radiation. Yet certain gases in the atmosphere – carbon dioxide, methane, water vapour – prevent that heat escaping into space. Instead, they absorb and re-emit it, so contributing to warming of the Earth's atmosphere. The notion of heat being trapped by this gaseous envelope has given the phenomenon its popular name of the 'greenhouse effect' (though heat retention in greenhouses actually works quite differently). As a natural process, it helps to maintain an even temperature in the atmosphere, but huge emissions of CO_2 and other gases by human activity in recent decades has magnified its effect. The debate continues about the long-term impact this will have on climate.

SCIENTIST SIBLINGS

William Herschel's sister Caroline (1750–1848) was a gifted astronomer in her own right, and assisted her brother throughout his career. She discovered five comets in the late 1700s and in 1828 was awarded a gold medal for her work by the Royal Astronomical Society.

Heat signatures
Thermal imaging can measure the heat given off by an inanimate object such as a house (above) or a living creature such as a lion (left) by detecting infrared radiation. In thermographic cameras, the coldest areas (0°C or below) show up as blue, grading through green and yellow zones to the warmest (24.1°C or higher), which appear red.

Continuous electric current

Inspired by his fellow Italian Luigi Galvani's work on bioelectricity, physicist Alessandro Volta set about creating the first electric battery. His groundbreaking invention was the first source of a steady electric current and it opened up a whole new field of scientific investigation.

Electrical convulsions
A detail from a contemporary print showing Galvani's experiments with frogs' legs (right).

In a letter to the British botanist Sir Joseph Banks, dated 20 March, 1800, Alessandro Volta described a device for generating electricity at will. Banks later read Volta's letter to the Royal Society in London. Thus began a new era in the history of science. Up until then, no-one had succeeded in generating a continuous electric current – the most they had managed to produce were brief electrical discharges using capacitors, which were of little use for serious scientific experimentation. Yet electricity was the fascinating new force of the age – a few years before his discovery Volta, then a student in Como, had written a Latin poem in praise of it – and scientists were keen to understand its secrets

THE ELECTRIC RAY

In his letter to Joseph Banks, Volta likened his electric pile to a permanent Leyden jar or the stinging organ of the electric ray. He even suggested calling his device an 'artificial electric organ'. The electric ray (below) is a cartilaginous fish that can emit an electrical discharge of up to 220 volts to defend itself against attack or to stun its prey. Volta knew about these fish; since ancient times, doctors had used them to alleviate gout and headaches or to numb the pain of childbirth. We now know that their electrical charge is produced by a mechanism that is present in all living cells. The passage of ions (electrically charged atoms) through channels in a cell membrane generates minute electric currents. The cells in electric organs have far more channels than normal and so can emit far more powerful charges.

From frogs' legs to the electric pile

In 1775 Volta was appointed professor of physics at the University of Pavia. His initial interest in electricity was sparked by his work on improving a machine that produced a static electric charge. This device, which Volta christened the 'electrophorus', was invented in 1762–6 by a Swedish physicist, Johan Carl Wilcke, and was in essence a refined version of the Leyden jar. (Discovered by chance 20 years earlier by German scientist Ewald von Kleist, the Leyden jar was a form of capacitor – an apparatus for storing electrostatic charges.)

In 1791 Luigi Galvani, an anatomist from Bologna, published a commentary on the effect of electricity on muscular motion. This detailed the remarkable results of a series of experiments he had conducted in 1786. He had taken dismembered legs from freshly killed frogs, bound them together with a copper strip and hooked them onto an iron balustrade around his terrace. When the copper touched the iron, the frogs' legs twitched convulsively, as if they had been connected to a Leyden jar.

Galvani called the phenomenon 'animal electricity', because he thought it came from a 'life force' within the frog tissue. He failed to realise that the muscular spasms actually resulted from the difference in potential between the two metals, the copper and the iron, and had little to do with the frog. Volta worked out that the frogs' legs were acting as both a detector and a conductor of the electricity between the metal braces. For him, Galvani's work proved the existence not of 'animal' but of 'metallic' electricity – an electric charge generated by the contact of two different metals. To test this hypothesis, he took a beaker of brine, into which he dipped one strip of zinc and one of copper. This simple contact produced an electric charge.

Over the next few years, Europe's scientific community was split between followers of Volta and 'Galvanists'. The controversy came to an end in 1797, when France invaded Italy and Galvani was sacked from his post for refusing to swear allegiance to the new regime. He died the following year in obscurity. Meanwhile Volta, who did pledge loyalty to Napoleon, conceived an apparatus in which paired discs of different metals, interspersed with pads of cardboard or cloth moistened with brine or sulphuric acid, were stacked one on top of the other – hence the term 'pile'. The two ends were connected with a metal wire so that the 'electrical fluid might flow uninterrupted'. Thus was born the voltaic pile, and from it the principle of the electric battery: two electrodes of different metals separated by an electrolyte (brine, acid or living tissue). Galvani had not been entirely wrong: 'animal electricity' was later proved to exist within living tissue, in the

THE BAGHDAD BATTERY

Iraq's National Museum houses a set of curious artefacts: small terracotta jars, each containing a copper cylinder wrapped around an iron rod. At the top, the iron rod is isolated from the copper by a bitumen plug, which seals the jar tightly. In 1938, Austrian archaeologist Wilhelm König dated these jars to the 3rd century BC and suggested they may have been voltaic cells for plating gold onto silver objects. In the late 1940s, tests by the General Electric High Voltage Laboratory in the USA seemed to bear out this theory. On the other hand, the absence of wires or conductors argues against the hypothesis, and the gold-cyanide salts necessary for gold electroplating, which do not occur in nature, were only known to have been available from the Middle Ages.

Emperor and subject
A colour print of a 19th-century engraving shows Volta (below) explaining the principle of his electric pile (far left) to Napoleon, who invited the scientist to Paris in 1801. Volta's work was instrumental in inaugurating the new science of electrochemistry.

Investigating electrolysis
An illustration of Michael Faraday conducting his experiments into electrolysis in 1834. Faraday used a voltmeter to show how much electricity was used during the operation. The scales shown in this engraving were to confirm that the quantity of tin deposited on the electrode was directly proportional to the quantity of electricity at that electrode and this finding became Faraday's first law.

form of minute electrical impulses generated in the nervous and muscular cells. Nor was Volta entirely right: electricity was not produced by the contact between two metals but by the chemical reactions this set off.

Intensive research

In May 1800 two British chemists, Anthony Carlisle and William Nicholson, used a voltaic pile to conduct the first conclusive electrolysis of water into hydrogen and oxygen. Thereafter, electrolysis became a fundamental process in chemistry, leading to the discovery of new elements. A friend of Volta's, Luigi Brugnatelli, was quick to realise that electrolysis could be used for electroplating metals. In 1834 Michael Faraday formulated his laws of electrolysis and coined the terms 'cations' and

PRESERVED FOR POSTERITY

The first International Electrical Congress, which met in Paris in 1881, reached agreement on the standard definitions and names of electrical units. In homage to Volta, the unit of electromotive force (or of potential difference) was named the volt. Galvani was not accorded the honour of having an SI unit named after him, but he did give his name to a device invented in 1820 that measured the intensity of an electric current – the galvanometer. These two great Italian scholars are also commemorated on the Moon, in the names of two neighbouring craters.

Early electrical lab equipment
A Nobili astatic galvanometer (right) attached to a thermoelectric battery in a detail from a 19th-century engraving.

'anions' to describe the electrically charged particles that move towards the two electrodes, and thereby create the electric current.

In 1820 the Danish physicist and chemist Hans Christian Oersted found that a wire attached to a voltaic pile moved the needle of a magnetic compass, thus demonstrating the link between electricity and magnetism. Oersted's work prompted intensive research into electromagnetism throughout Europe. Within a few decades, its fundamental laws had been formulated and key pieces of apparatus like the magneto, the electric motor and the galvanometer were already a reality.

Ongoing improvements

The voltaic pile suffered from several inherent flaws: it was fragile; it was dangerous to handle due to the acid; and its power output was quickly impaired by hydrogen forming on the copper electrode. New inventions, notably the electric telegraph, demanded a continuous and reliable current. In 1829 French chemist Antoine-César Becquerel devised an acid-alkali cell comprising two electrodes (conductive strips), one of zinc in sulphuric acid and the other of copper in copper sulphate, separated by a porous membrane made from a cow's intestine. This was the world's first constant-current electric cell, and could supply current for up to an hour. But by any standards, it was highly impractical.

GIANT BATTERIES

In his original electric pile, Volta combined 20 separate elements but was still only able to generate a current about as strong as that of an 'extremely sluggish electric ray'. Far more power was needed for electrolysis, prompting the construction of giant batteries. In 1813 Napoleon had a huge battery of 600 paired copper and zinc plates built, occupying an area of 54 square metres at the Polytechnic College in Paris. That same year, Humphry Davy and William Hyde Wollaston put together an even more massive battery containing 2,000 pairs of plates. Such monster batteries could generate tens of kilowatts of power.

Subterranean generator
Davy and Wollaston constructing their giant battery in the basement of the Royal Society in London in 1813.

In 1836 a British chemist, John Daniell, replaced the cow's intestine with an unglazed earthenware pot containing zinc sulphate (in place of the sulphuric acid). This was the first truly practical and reliable battery. Then, in 1859, the French physicist Gaston Planté invented the lead-acid storage battery, which could be recharged repeatedly; this type of battery is still used for cars. Eight years later, Georges Leclanché created a battery combining an electrolyte of ammonium chloride with electrodes of zinc (negative) and crushed manganese dioxide (positive). The Leclanché cell, widely used in telegraphy, was the forerunner of the modern dry-cell battery. Finally, the first true dry-cell battery, in which the liquid electrolyte was replaced by a gel, was patented by the German Carl Gassner in 1887. This key innovation made batteries transportable for the first time. Since then, batteries have changed little right up to the present day.

Push start
In a car battery, a chemical reaction between an electrolyte solution and the battery's lead and lead oxide plates releases electrons that produce an electric current. This in turn activates the starter motor, which turns over the car's engine. When a battery goes flat, as in this old Citroën 2CV, one answer is to use human muscle power to give the car a push start.

NAPOLEON'S SCIENTIFIC AND ARTISTIC COMMISSION

State-sponsored science

Flushed with the extraordinary success of the team of archaeologists and historians he took with him on his Egyptian campaign, Napoleon placed French scholars and scientific institutions at the service of the state. In doing so, he was continuing a tradition of public research in France that began with the French Revolution.

Historic address
The lid of an ivory and metal box depicts Napoleon speaking to his troops on campaign in Egypt (top right). The inscription translates: 'From the heights of these monuments, forty centuries of history look down upon you.' The mathematician Pierre-Simon Laplace (right) was an avid supporter of Napoleon and his campaign in Egypt.

During the Reign of Terror, many of France's foremost scholars were sent to the guillotine. The mathematician and astronomer Jean-Sylvain Bailly was executed in 1793 in his capacity as mayor of Paris, while Lavoisier's role as a tax-collecting *fermier-général* sealed his fate in 1794. Yet as Republican France increasingly came under threat from its worried European neighbours, it could ill afford to purge its leading scientists.

In April 1793 the Committee of Public Safety, with the chemist Guyton de Morveau presiding, established a 'Commission of four citizens versed in chemistry and mechanics charged with researching and testing new systems of defence'. Although this failed to save Lavoisier, other scientists took up his 1788 research, conducted with Claude Louis Berthollet, into potassium muriate gunpowder. Their efforts lead to a new naval shell developed in the laboratory of the National Balloon School at Meudon, founded in 1796 to promote the military use of balloons. Two years earlier the National Convention had founded the Academy of Arts and Crafts, dedicated to technical education and developing inventions, and the Central School of Public Works (renamed the Polytechnic School in 1795) under the direction of mathematician Gaspard Monge. In 1795, Lavoisier's name was officially rehabilitated, and the Academy of Sciences, abolished in 1793, was reformed. In 1797 it elected to its ranks a young general by the name of Bonaparte – a keen amateur mathematician – in recognition of his successes in Italy.

THE FRENCH NEWTON

Pierre-Simon Laplace (1749-1827) was professor of mathematics at the French military academy when he interviewed a young cadet named Napoleon for the artillery. Upon becoming First Consul, Bonaparte made Laplace his minister of the interior. Sadly, the academic was such a poor administrator he lasted only six weeks in the post. But he was a brilliant polymath. In 1796 he advanced the nebular hypothesis for the origins of the solar system, which held that all celestial bodies were formed from a cloud of gas and dust. Laplace also postulated the existence of black holes. In physics, he helped Lavoisier with his calorimetric measurements and formulated the laws of electromagnetism. His greatest contribution to mathematics was his work on probability theory. After the Bourbon restoration, he was made a marquis.

An Enlightenment agenda

In 1798 the *Directoire*, the new ruling council of France, dispatched this ambitious commander on a mission to invade Egypt. The political aim

was to cut Britain's supply line to India and tie her down on a front far from home. On 19 May, 400 ships laden with 400,000 troops, 10,000 sailors and 300 women set sail from Toulon. Also in the party – and with a more lofty aim to 'foster and propagate the ideals of the Enlightenment' – were 150 artists and scholars, the cream of France's mathematicians, engineers, naturalists, geographers, draughtsmen, interpreters and printers. Their names are a roll-call of the country's surviving intellectual élite of the time: Monge, Denon, Berthollet, Geoffroy, Saint-Hilaire, Cuvier, Laplace and Conté (inventor of the modern lead pencil). The plan was for the scholars to study Egypt's rich past through its archaeology, as well as its natural resources and people, with an eye to eventual colonisation.

On arrival in Cairo in July 1798, Napoleon immediately founded the Egyptian Institute. This was organised into four sections: mathematics; physics and natural history; political economy; and literature and the arts. Its members accompanied French troops on their forays into Upper Egypt, taking measurements and making sketches.

The pharaohs of the 12th Dynasty (*c*2000–1800 BC) had cut a canal linking the Nile to the Red Sea, but this had been destroyed in the eighth century AD by the

Soldiers and scholars
Leon Cogniet's Egyptian Expedition under the Orders of Bonaparte *(1821) shows Napoleon, Kléber and archaeologist Dominique Vivant sheltering under a canopy from the fierce desert sun.*

Linguistic touchstone
The Rosetta Stone is one of the most famous exhibits in the British Museum in London.

DECIPHERING THE PAST

Uncovered by Captain Pierre-François Bouchard in Rosetta (Rashid), the Rosetta Stone is a stela inscribed with a 2nd-century-BC decree proclaiming Ptolemy V ruler of Egypt. The reason it has become so renowned is that it is in three languages – hieroglyphics, Demotic and Classical Greek – and this gave scholars a key to unlock the secret of Egyptian hieroglyphics. The stone was dispatched to Alexandria, where a copy of the text was made and sent on to Paris. Eventually the stone ended up in the British Museum, where Thomas Young deciphered many of the hieroglyphs, but reliance on a fragmentary work on Demotic characters by Swedish scholar Johan Åkerblad led to a number of errors. The French philologist Jean-François Champollion finally cracked the hieroglyphic code in 1822.

Egyptian enigma
A copperplate engraving of the Sphinx at Giza, from Travels in Lower and Upper Egypt *(1802) by Dominique Vivant, Count Denon.*

INSPIRING A FASCINATION WITH EGYPT

In 1799 General Kléber gathered together all the research undertaken by the Egyptian Institute for publication in a single work. It took until 1808 for the first volume of *Description of Egypt* to appear, by personal order of Napoleon, and publication continued until 1826. It comprised nine volumes of text, a comprehensive atlas of Egypt (classified as a military secret until 1814) and ten volumes of plates – five devoted to ancient Egypt, two to modern Egypt and three to the country's natural history. The work sparked enormous enthusiasm for all things Egyptian throughout Europe. The 'Egyptian Style' became fashionable in the decorative arts and architecture, and new museums sprang up dedicated to Egyptology, including the Museo Egizio in Turin and the Ägyptisches Museum in Berlin.

Ancient calendar
A plate in the Description of Egypt *reproduces the lunisolar zodiacal calendar from the ceiling of the temple of Hathor at Denderah.*

Abbasids. Bonaparte now dispatched a civil engineering mission under Jacques-Marie Le Père to survey the Isthmus of Suez with a view to excavating a new canal. Le Père's team wrongly identified a 10-metre difference in height between the Mediterranean and Red seas. The miscalculation was forgiveable – all their instruments had been lost when Nelson destroyed the French fleet at the Battle of the Nile at Aboukir Bay on 1 August, 1798.

Napoleon tried – and failed – to invade Syria, then returned to France in August 1799, leaving the Egyptian mission under the command of General Kléber. The scientists continued their work, producing reports on a wide range of topics, from a treatise on mirages by Mong to a report on Egypt's natron deposits (a natural form of soda ash) by Berthollet. Although the Egyptian campaign ended in military disaster (Kléber was assassinated in 1800 and the French army retreated the following year), the scientific mission had great propaganda value. Its legacy also inspired the Egyptian viceroy, Mehmet Ali, to institute modernising reforms. These were well underway by the time Ferdinand de Lesseps began cutting the Suez Canal in 1859.

Shells and rockets

Shortly after his return to Paris, Napoleon overthrew the *Directoire* and installed himself as First Consul. He then set about root-and-branch reform of government, which included strengthening the link between the corridors of power and the academic world. On crowning himself Emperor, he made the Polytechnic School a military institution and its graduates would play key roles in scientific and political life both during his reign and the Bourbon restoration that came after. Behind the military character of the institution lay a technocratic model of public research that was to leave its mark not just on higher education in France but also in the wider world.

In 1810 Napoleon formed a commission on gunpowder production, which led to Jean-Siméon Champy perfecting the 'round-powder' process for the manufacture of safer, granulated gunpowder. Another state commission, chaired by Monge, looked into the use of incendiary rockets in an effort to keep pace with British advances in this field under William Congreve. French rockets were deployed against Spanish towns during the Peninsular War.

In 1814 Polytechnic students were active in the defence of Paris. Napoleon would soon be toppled, but France's future scientists felt more than ever bound up with the fate of the nation.

Egyptian style
An ornate bedside table in the Egyptian style designed by the renowned French furniture makers, the Jacob brothers, and the architect Louis-Martin Berthault, who was ennobled by the Empress Josephine in 1805.

Imperial patronage
A visit by Napoleon to the Louvre is recalled in a painting of 1825 by Louis-Charles Auguste Couder (left). The Louvre had lost its function as a royal palace when Louis XIV moved to Versailles in 1674. After the Revolution the National Assembly decreed that it become a public museum and it opened its doors in August 1793.

NAPOLEON'S HANDY MAN

Napoleon set up the headquarters of his Egyptian scientific and artistic commission in the former palace of Hassan-Kashif on the outskirts of Cairo. The inventor and balloonist Nicholas Conté (1755–1805) was put in charge of the institute's workshops. After most of the expedition's scientific apparatus went down with the French warships sunk at the Battle of the Nile – and an uprising by the people of Cairo in October 1798 that destroyed further equipment – Conté's skills were much in demand. The workshop turned out a huge variety of implements, including windmills, clocks, parts for printing presses, and even some sophisticated surveying tools. Napoleon greatly valued Conté, calling him 'a universal man of taste, understanding and genius, capable of creating the arts of France in the Arabian desert'.

MORPHINE – 1806

Natural painkilling – at a price

The opium poppy has been known since ancient times as a plant with calming and painkilling medicinal properties, but it also has a long history of addiction and abuse. Morphine was first extracted from the poppy by a German apprentice pharmacist, Friedrich Wilhelm Sertürner. The discovery gave the medical profession a powerful new means of combating pain.

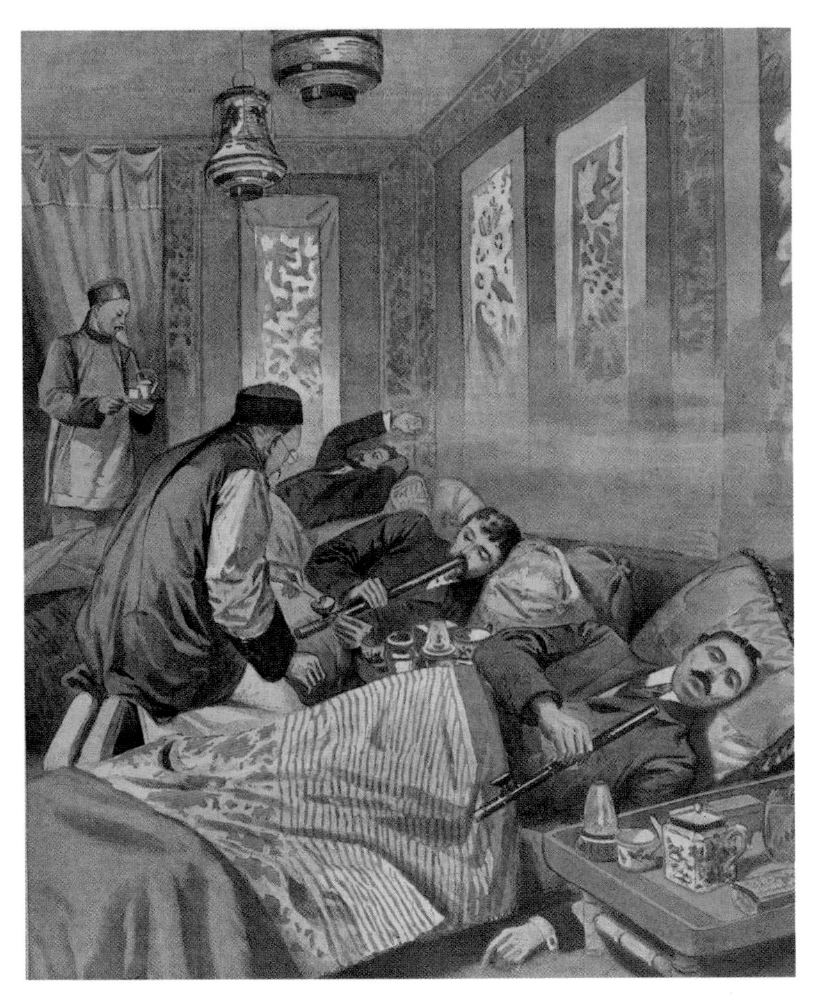

Narcotic haze
An opium den in Paris, from a magazine illustration of 1907. By 1916, there were thought to be around 1,200 such clandestine establishments in the city.

Down the ages, Western medicine had largely neglected the question of pain relief. Christian teaching held it a virtue to suffer in silence, while the ancient Greek physician Hippocrates regarded pain as a useful 'watchdog' for alerting a person to illness. There were folk remedies to hand. Peoples from the ancient Sumerians, Greeks and Romans to Native Americans knew the secret of infusing the bark of the willow tree to obtain relief from aches and pains. Willow was later found to contain salicylic acid, which was synthesised into Aspirin. But such remedies were of no avail in extreme cases, such as amputations.

In the early 19th century the medical world began to reassess a plant known since ancient times for its narcotic and analgesic properties: the opium poppy (*Papaver somniferum*). Derived from the latex tapped from the plant's seed heads, opium appeared to lessen acute pain considerably, but preparations containing it had some highly unpredictable effects. The aim was to find a way of isolating the active painkilling ingredient. French and German chemists set to work on the problem, armed

Popular poison
Laudanum was famously associated with the Romantic poets such as Coleridge and Shelley, but it was widely prescribed for all manner of ailments in the 19th century and was the drug of choice among all ranks, from the working class to Queen Victoria.

TIME-HONOURED PAINKILLER

The painkilling properties of the latex from poppies was known to the Sumerians, Egyptians, Greeks and Romans. Scholars think that the 'Nepenthe' or 'drug of forgetfulness' mentioned in Book 4 of Homer's *Odyssey* was an opium-based preparation. In the 1500s, the Swiss alchemist Paracelsus introduced the drug into Western medicine. The following century, English physician Thomas Sydenham was the first to describe its effects; he also concocted a more effective tincture of opium known as laudanum.

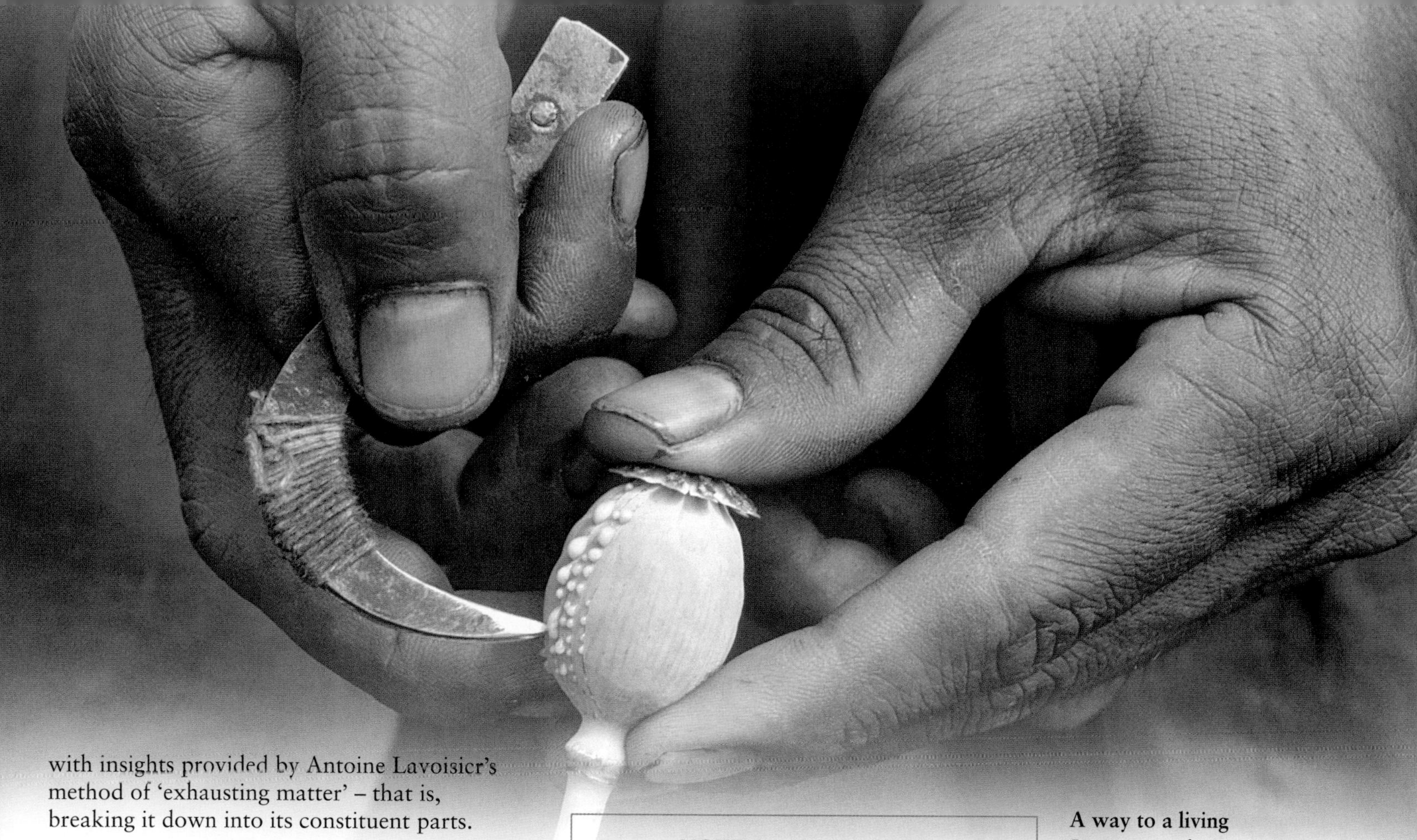

with insights provided by Antoine Lavoisier's method of 'exhausting matter' – that is, breaking it down into its constituent parts.

Dedicated research

In his Paris dispensary, Jean-François Derosne steeped opium poppy heads in distilled water, then let the solution evaporate before scraping out the resulting brown crystals and adding water to them. He thought he had succeeded in isolating the plant's active ingredient, which he christened 'Derosne's salt' (later known as narcotise), and published his results in 1803. The next year Armand Seguin, a former pupil of Lavoisier, refuted Derosne's findings in a paper describing his own procedure. Without realising it, Seguin had in all likelihood discovered morphine, but before he had a chance to pursue his work further he was arrested on charges of embezzlement.

The breakthrough was made in Germany by Friedrich Wilhelm Sertürner (1783-1841) of Paderborn. At the age of 15 Sertürner had followed in his father's footsteps and taken up chemistry. After securing a position as a pharmacist's assistant in 1803, he persuaded his employer to let him study opium, which fascinated him, paying for the raw materials out of his own pocket. Two years later, he announced that he had discovered a substance he called 'meconic acid'. He administered this to an injured dog, but found it had no effect. By contrast, another substance he had isolated, a vegetable alkali, had a powerful analgesic effect. At first, his research made little impact, and Derosne's salt continued to be used by physicians. Sertürner wrote at the time: 'Our results [his and Derosne's] were so different and contradictory that no-one could get to the truth of the matter. My study, which I'd jotted down quickly, received scant attention … and many people claimed to have repeated my experiments without success…'. But his efforts were eventually rewarded. In 1817 he proved that his alkali was indeed the active ingredient of opium. He called the white crystalline substance he obtained 'morphium' – after Morpheus, the Greek god of sleep.

Powerful effects

Morphine was the very first vegetable alkali, or alkaloid (a nitrogen-containing organic alkali), to be isolated by chemists. Its discovery opened up a whole new field of study for Sertürner and his contemporaries, as they strove to unlock the mysteries of drugs extracted from animals and plants. While experimenting on himself and three young men, Sertürner found that morphine in large

USEFUL ALKALOIDS

The discovery of opium alkaloid morphine revolutionised pharmacology. Thereafter, chemists applied the same techniques to extract from plants several other alkaloid substances including strychnine (1818), quinine (1820), nicotine (1828), codeine (1832) and atropine (1833). Most recently, in the 1980s, the anti-cancer drug docetaxel was derived from the needles of the European yew tree (*Taxus baccata*).

A way to a living
Latex oozes from an opium poppy seedhead in the hands of a member of the Lisu people who inhabit the mountainous region on the borders of Thailand, China, Burma and India. The opium poppy is often cultivated as a cash crop to subsidise the meagre living of subsistence farmers.

Bloody conflict
Liberal use of morphine in the American Civil War (1861-5) and the Franco-Prussian War (1870-1) confirmed its efficacy, but resulted in the 'soldier's disease' of morphine addiction. The painting above is of the Battle of Chancellorsville (1863).

Bone saw
A saw used in amputations in the 19th century (above). Before the advent of antiseptics, amputation was often the only way to stop infection spreading.

doses could be dangerously toxic. The weakest of his three patients almost died, while he lapsed into a state of semi-consciousness, punctuated by violent bouts of vomiting.

In 1827 the German chemist Heinrich Emanuel Merck began producing morphine on an industrial scale. Its spread was promoted by the invention of the hypodermic needle in 1853. Yet the use of morphine in battlefield surgery gave rise to the first widespread drug addiction in Western history. Famous figures of the period were among those affected, including Bismarck and Sigmund Freud. Sertürner died a morphine addict in 1841.

The dangers of morphine were finally acknowledged in 1868 when Britain declared it a 'controlled substance'. The USA followed suit in 1914. In 1898 the Bayer chemical company in Germany synthesised heroin from opium and marketed it as a nonaddictive morphine substitute, but addiction problems soon arose. The structural chemical formula of morphine was determined in 1925. A safe synthetic derivate was first patented in 1952.

Today, the analgesic paracetamol or anti-inflammatories like ibuprofen are usually taken for aches and pains. In more acute cases, weak morphine derivatives, such as the opiate codeine, are prescribed at first, followed by morphine itself or a stronger derivative if the pain persists. In palliative care, such as for cancer sufferers, it is not uncommon for patients to self-administer the drug by means of a morphine pump.

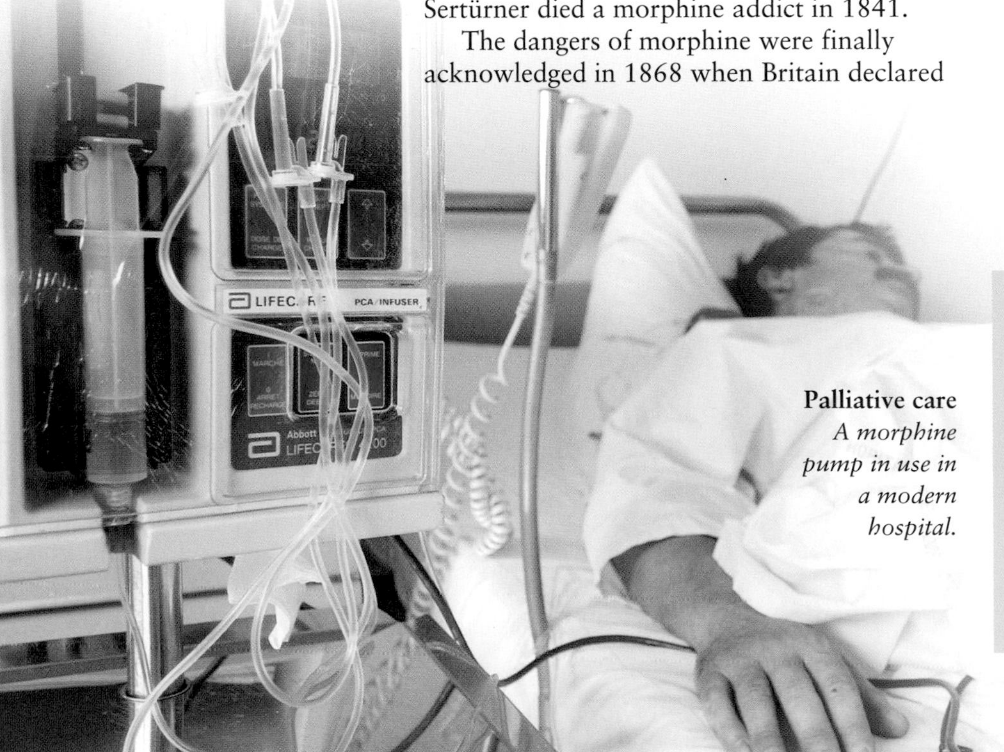

Palliative care
A morphine pump in use in a modern hospital.

STRIKING RESEMBLANCE

Research in the United States in the early 1970s revealed that morphine fixes on specific receptors on the surface of neurons in the brain. In fact, the drug's structure is strikingly similar to that of analgesic substances known as endorphins that are produced by the brain itself. Both the drug's painkilling properties and its undesirable side-effects are caused by its action on the brain itself.

The chromatic trumpet *c*1800

A basic trumpet made from bones, horns or shells is one of the earliest instruments known to humankind. Metal trumpets first appeared in ancient China and India, and later became popular in Greece, Rome and Egypt. A silver trumpet was found among the treasures in Tutankhamun's tomb. The natural trumpet – comprising a mouthpiece, tube and bell – was the main brass instrument played in the Middle Ages. The modern form of trumpet, with the tube bent into three parallel branches, is thought to have originated in the Low Countries in the late 1400s.

By the early 1800s composers were beginning to find the harsh tone and limited range of the natural trumpet restricting. The technique of hand-stopping – placing the hand over the bell to deaden the sound – was introduced in 1775, but this was not enough to mask the instrument's perceived deficiencies.

Military honour
An 18th-century natural trumpet (left), presented by Napoleon to Citizen Norberg for conspicuous bravery at the Battle of Marengo in 1800 during the French campaign in Italy.

From keys to pistons

Trumpet-makers were keen to make instruments capable of reproducing all the notes of the chromatic scale in semitones. The addition of a system of keys, as on a transverse flute, enjoyed brief popularity in Italy and Austria. In 1803, the *Trumpet Concerto in E Minor* by Johann Nepomuk Hummel helped to popularise the keyed trumpet, but the resonance of the instrument was uneven and it soon fell out of favour. English instrument-makers John Hyde and John August Köhler produced modern versions of the Renaissance slide trumpet, but encountered similar problems.

In *c*1810, the rotary valve and the piston valve were invented by Berlin trumpet-makers Friedrich Blühmel and Heinrich Stölzel respectively. Once pressed, the valves opened up a supplementary windway into which the air was diverted, producing a low, resonant sound. In 1818, the pair jointly patented the three-piston trumpet. Later improved by Frenchman François Périnet, this became the basis of the modern three or four-valve chromatic trumpets that deliver a crisp and even tone.

SATCHMO, DIZZY AND MILES

The trumpet came into its own in the hands of the great 20th-century jazz trumpeters, such as Louis Armstrong ('Satchmo'), Dizzy Gillespie, Chet Baker and Miles Davis (below). Towards the end of the century, classical trumpet found a new popularity through players like Maurice André, who rediscovered Baroque compositions and forgotten classical pieces.

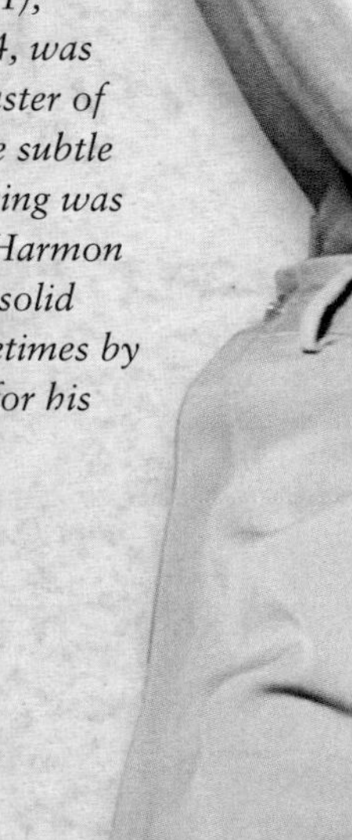

Blow that horn
Miles Davis (1926–91), pictured here in 1954, was an acknowledged master of the jazz trumpet. The subtle resonance of his playing was achieved by using a Harmon mute (a mute with a solid ring of cork) or sometimes by substituting a bugle for his usual instrument.

Electroplating 1805

Royal visit
An engraving from the Illustrated London News, *show Alexandra, Princess of Wales, wife of the future Edward VII, dipping a vase into an electroplating bath during a visit to a Birmingham factory in 1874.*

Work of art
This ornate gold-plated goblet (right), decorated with a frieze of cherubs, was produced by the Barbedienne foundry in Paris in c1875.

In 1805 Luigi Brugnatelli, an Italian chemist, discovered that if a silver coin was attached to the cathode of Alessandro Volta's electric pile then dipped in a solution of gold salts, the coin emerged coated with gold. This was the very first instance of electroplating – the electro-chemical process of depositing metal on objects. Brugnatelli demonstrated his discovery to Napoleon, but the French emperor was unimpressed and refused him the right to exploit his work commercially.

Later, other scientists unearthed Brugnatelli's only paper on the subject, published in an obscure Belgian journal, and revived his process. In 1837 the Prussian engineer Moritz Hermann von Jacobi developed electroplating with silver, nickel and chrome – a method he called 'galvanoplasty' after electrical pioneer Luigi Galvani. Three years later, a Birmingham doctor named John Wright found that potassium cyanide was an ideal electrolyte for plating silver and gold. Patents were taken out by Henry and George Elkington in England and Ernst von Siemens in Germany. Used at first for decoration or protection, electroplating has since found new applications, notably in the manufacture of computer chips.

WEDGWOOD'S
HIGHLY IMPROVED,
PATENT MANIFOLD
WRITER,
FOR
COPYING LETTERS,
INVOICES, DRAWINGS, PLANS, &c.

Carbon paper 1806

James Watt's press-copier (see below) failed to catch on, so in the early 19th century the only way of producing a duplicate of a document was to copy it out by hand. In 1806 the English inventor Ralph Wedgwood came up with carbon paper: a piece of paper coated with coloured wax, which was placed between two sheets of writing paper. The pressure applied in writing on the top sheet went through the carbon paper transferring wax onto the bottom sheet. It required a neat and rather heavy hand for the copy to be legible, so carbon paper did not really take off until the advent of the first typewriters and thereafter the ballpoint pen.

Labour-saver
An 1834 advertisement for Wedgwood's 'Patentmanifold Writer'. The target market included solicitors, bankers and tradesmen.

WATT'S PROTOTYPE PHOTOCOPIER

In 1780 James Watt patented his 'press-copier', which was, in effect, a rudimentary photocopier. The original document, written in a mix of ink and gum arabic, was laid face-down on damp tissue paper. The machine pressed the two sheets firmly together, transferring a copy of the writing to the tissue. Although the copied text came out back-to-front, it was legible from the reverse side through the thin tissue paper.

ROCKETS – 1806

First steps into space

Rocket technology had its origins in China in around the 12th century AD, but the birth of modern rocketry can be dated to 1806, when the British bombarded the French port of Boulogne with self-propelled projectiles invented by Colonel William Congreve. Sophisticated modern rockets have dual roles as powerful weapons of destruction and as launch vehicles for space exploration.

'I have succeeded in making 32lb [14kg] rockets containing as much explosive as can be packed into a spherical carcass 10 inches [250mm] in diameter and with an average range of 3,000 yards [2,700m]'. Thus ran a progress report by Colonel William Congreve to the British government in March 1806. He was working on the invention that was later to bear his name: the Congreve rocket. He confidently predicted that his brand-new weapon of war might one day even displace the cannon. In the same letter, he sought leave to give a 'live-fire' demonstration, which duly took place on 8 October that year.

In the presence of George III, a Royal Navy flotilla armed with prototypes of the Congreve rocket dropped anchor more than a mile off the French port of Boulogne, where Napoleon was assembling an invasion fleet against England. The rockets were fired at the French ships, but they were blown off course by a stiff breeze and fell on Boulogne itself, setting the town ablaze and sending waves of alarm through Europe. The following year the British bombarded Copenhagen with 25,000 incendiary projectiles, starting a spectacular conflagration that virtually razed the Danish capital to the ground. In 1813 it was the turn of Leipzig and Danzig to deploy the new weapon – this time against Napoleon's besieging armies.

Chinese arrows

For all the mayhem they unleashed, Congreve's rockets were really just modern adaptations of an older technology originating in the Far East. The British had learned to their cost about the destructive power of rockets in 1789. During a bitterly fought campaign against the prince of Mysore Haidar Ali and his son Tipu Sultan, the Indian rocket corps wreaked havoc among the British cavalry. By then, the rocket

A NEW WORD FOR A NOVEL WEAPON

The first known mention of the word 'rocket' in Europe occurs in a report written in 1379 by an Italian named Muratori, who called the device a *rocchetta*, or 'small fuse'. In recounting an attack by the Paduan army on Mestre near Venice, Muratori describes the use of small incendiary projectiles, which probably resembled modern fireworks. Over the following centuries, the use of rockets as military weapons became increasingly common. In 1715, Tsar Peter the Great of Russia ordered construction of the world's first rocket factory in St Petersburg.

Raining fire from the sky
A Congreve rocket dating from 1817 (left). These projectiles comprised a metal cartridge filled either with incendiary material or with high explosives. The tail fins added stability in flight.

City in flames
Britain's victory in the Second Battle of Copenhagen in 1807 was secured with the help of a devastating Congreve rocket attack on the Danish fleet and the city.

had been in use in China for at least 500 years. In 1232, for example, Jin forces in Kaifeng, the capital of Henan province, routed a besieging Mongol army with rockets.

In 1280 the Arab scholar Hassan al-Rammah wrote a treatise entitled *The Book of Fighting on Horseback and with War Engines*, describing in great detail the manufacture of gunpowder and rockets. In common with other Arab commentators, he called the rockets 'Chinese Arrows'. Over time, improvements were made to every aspect of these devices – to the shape of the projectile, the material for the casing, the mode of propulsion and the guidance system.

Deadly battery
A case of Chinese 'fire arrows', primed for firing.

Unprecedented range

Yet the real innovation that the Indian Mughal armourers – and after them the British – brought to rocketry was the use of metal cases, which replaced the tubes of compressed and glued paper used in the Chinese devices. The advantage was decisive: this robust new casing allowed Tipu Sultan's rocket lancers to pack their weapons with gunpowder, safe in the knowledge that they would not explode prematurely during the launch or in flight. The greater thrust available from metal rockets also

FROM EAST TO WEST

The Mongols learned quickly from the rout of their cavalry by Chinese 'fire arrows' at the siege of Kaifeng in 1232. Less than a decade later, in 1241, they deployed rockets against European forces at the Battle of Legnica in Silesia. In 1288 Almohad forces used rockets to attack the city of Valencia as they fought for control of Al-Andalus in Spain. Use of such rockets then spread fast – *Bellifortis*, a 1405 treatise on war by Konrad von Eichstädt, mentions three different types.

MORE THAN A ROCKET-MAN

William Congreve (1772–1828) is remembered today for his rocket weapons, but he was a prolific inventor in many different fields. His diverse portfolio included a colour printing process, a perpetual motion machine, a new type of steam engine, a canal lock and sluice operated by hydropneumatic power, an unforgeable banknote, and a system for absorbing smoke. Back on the deadly side, he also came up with a parachute attachment that enabled rockets to float down onto their targets and an exploding rocket harpoon for killing whales. Perhaps the most unusual memorial to Congreve resulted from an unsuccessful British attack in the War of 1812 (which lasted until 1815) against the USA – the bombardment of Fort McHenry in Baltimore in 1814. This action prompted Francis Scott Key, author of the American national anthem *The Star-Spangled Banner*, to include in the first stanza a reference to 'the rockets' red glare'.

Deadly streamlining
William Congreve got the basic idea for his rocket from Chinese and Indian weapons, but he improved them with an aerodynamic shape, which increased their speed and stability in flight.

Final showdown
When the British stormed the citadel at Seringapatam in 1799 (above), they were opposed by the formidable rocketeers of Tipu Sultan (right). A decisive moment came when a British shot hit a magazine of rockets inside the fort, causing a huge explosion.

dramatically increased their range, propelling them for over a mile – a distance unheard of for paper rockets. To compensate for the lack of accuracy, attackers simply saturated the target with a volley of projectiles.

Indian rockets were basically metal cylinders, closed at one end, measuring some 20cm in length, with a diameter varying from 3 to 7.5cm. When the propellant was ignited, it produced a large volume of gas that was forced violently out of an exhaust nozzle at the base of the rocket. The rockets were lashed to bamboo guide-sticks over a metre in length, which enabled the weapon to be pointed in the right direction before firing. When the rocket landed, the shock of the impact caused the unspent gunpowder in the tube to explode.

Yet despite their undoubted destructive power, Tipu Sultan's rockets still lacked one vital feature of modern rockets. It would fall to William Congreve to put the finishing touches to this fearsome technology.

Passing the baton

British forces collected spent rocket cases from battlefields at Seringapatam and elsewhere and took them back to London for analysis. In 1801 William Congreve was working in the laboratories at the Woolwich Royal Arsenal, where his father had formerly been Comptroller. One of the functions of the Arsenal was to manufacture fireworks 'as well for war as for triumph', and Congreve set to work unravelling the secrets of Tipu Sultan's rockets, soon coming up with an improved prototype of his own. He left an account of his efforts in his 1807 work entitled *A Concise Account on the Origin and Progress of the Rocket System*: 'It was in 1804 that I first hit

Arms for export
By the early 1800s, some 15 nations were using Congreve rockets. These illustrations, taken from Congreve's Rocket System *of 1814, show members of the British Rocket Corps loading and firing the weapon (top) and rockets being fired from Royal Navy ships (bottom).*

upon the notion that, since the thrust of rockets is contained within them and that they therefore cause no recoil reaction at the point from which they are fired, they could be used just as easily on sea as on dry land ... I knew that rockets had already been used in military operations in India but the size of these projectiles was negligible, while their range rarely exceeded 1,000 yards [900m].'

In view of the lightness and ease of firing of his new weapons system, Congreve neatly summarised rockets as being 'the soul of artillery without the body'. Congreve's work would have a major impact on later military technology and the aerospace industry.

Warheads and aerodynamics

Congreve's improvements were key to the subsequent development of the rocket. He began by inventing what is now known as the warhead: a compartment forming the nose of the projectile and containing an explosive charge (gunpowder in Congreve's day). The warhead is generally conical in shape, giving the rocket an aerodynamic profile. This not only helps to reduce friction as the rocket flies through the air, but also makes the ballistic trajectory more predictable.

Congreve then reduced the width of the rocket's base in order to increase the speed at which the combustion gases exited, and hence the thrust. In doing so, he effectively created the world's first rocket motor. His final development was to replace the single large exhaust nozzle with five smaller ones, which helped to stabilise the rocket in flight.

The first practical testing of a Congreve rocket took place at the Woolwich Arsenal in 1805. The prototype weighed 2.7kg and had a range of 1,800 metres. Ultimately, his rockets would increase to ten times that weight and fly for 2 miles. The technology did not remain British for long. Following the bombardment of Boulogne in 1806, French munitions experts set about developing rockets of their own.

The new weapons system was adopted by a growing number of nations around the world, and refinements came thick and fast. In 1844 William Hale introduced the stickless rocket, which lifted off solely by its own thrust and had angled fins fitted to the base to help it spin-stabilise in flight using its exhaust gases. In 1903 the Russian scientist Konstantin Tsiolkovsky published *The Investigation of Space by Means of Reactive Devices*, the first truly scientific work on the physics of rocket flight. Tsiolkovsky envisaged multistage rockets fuelled by liquid oxygen and hydrogen carrying humans into space. This idea was later revived by the French scientist Robert Esnault-Pelterie (1912), the American Robert Goddard (1920) and the German Hermann Oberth (1923). Oberth's and Goddard's work paved the way for Germany's V1 and V2 rockets during World War II, developed by Wernher von Braun, and for the American space programme, which peaked in 1969 with the Apollo 11 moon landing.

EARLY ROCKETEERS

Long before rockets took on their familiar modern form, with a conical warhead containing an explosive charge, aviation pioneers reputedly tried using the technology to get airborne. According to a Chinese account, probably apocryphal, an official named Wan Hu (born *c*AD 1500) attempted to become the first astronaut in history. He lashed a chair to two kites, attached 47 rockets around its base, then climbed on board and instructed his servants to light them all at once. In the ensuing blast, Wan Hu was atomised. Apparently blessed with better luck was the Turk Lagâri Hasan Çelebi, who in 1633 is said to have flown over the Bosporus on board a vehicle powered by seven rockets filled with 70kg of gunpowder. Lifting off from the Topkapi Palace in Istanbul, he flew for 200 seconds and reached a height of 300m before dropping unharmed into the water.

MULTISTAGE ROCKETS

Space flight only became a reality through the development of the multistage rocket. A multistage rocket – apparently a weapon used by the 14th-century Chinese navy – was described and illustrated in the work *Huolongjing* by Jiao Yu (1328–98), who called it a 'fire dragon issuing from the water'. In 1650 the Polish nobleman Kasimierz Siemienowicz gave a detailed technical account of how such a vehicle might be constructed, in his work *Artis Magnae Artilleriae* ('The Complete Art of Artillery'), which was widely read and translated into several European languages.

Rocket man
Following in the daredevil rocketeer tradition, Paul Reubens played human cannonball Pee-Wee Herman in the 1988 film Big Top Pee-Wee.

One giant leap
The Saturn V rocket carrying the Apollo 11 capsule blasts off from Cape Kennedy on 16 July, 1969. Five days later, at 10.56pm Eastern Standard Time, Neil Armstrong became the first man to walk on the Moon.

PARIS AFTER THE REVOLUTION

The transformation of a city

After the turmoil of the French Revolution, a new era dawned for Paris. Republican France, with the capital at its heart, came firmly under the control of Napoleon. As their city began to be transformed by pavements, gas street-lighting, new main roads, monuments and museums, Parisians began to indulge in gastronomic and artistic pleasures. A thriving economy drew in thousands from the countryside, and Paris gained a growing world reputation as a seat of learning.

In 1801 Paris had 547,000 inhabitants, making it the world's second most populous city after London. The Revolution had restored Paris to the pre-eminence it had lost in France when Louis XIV, the 'Sun King', had moved the court to Versailles. It was once again the decision-making heart of the country. Then, following the coup staged on 18 Brumaire of Year 7 of the revolutionary calendar – that is, 9–10 November, 1799 – Napoleon Bonaparte became the city's master.

Louis XIV had built broad concentric boulevards for citizens to stroll along. Under his successor, Louis XV, some remarkable buildings sprang up, notably the Corn Exchange (the

Showpiece plan
The imposing neoclassical Odéon theatre, from an engraving by Courvoisier (below), was opened in 1782. The whole city quarter around the theatre was also newly developed.

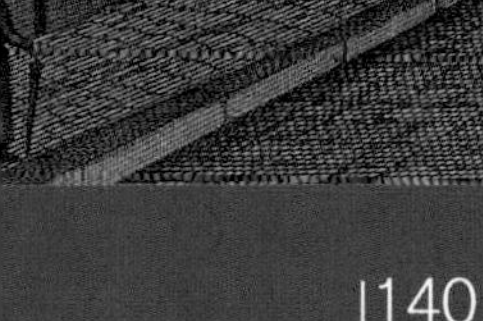

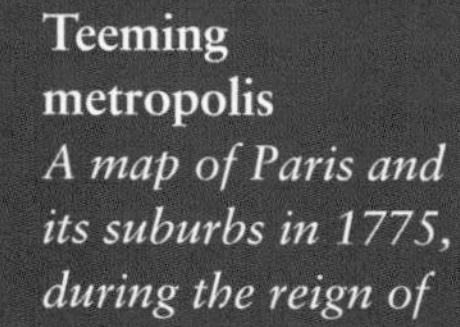

Teeming metropolis
A map of Paris and its suburbs in 1775, during the reign of Louis XVI.

modern Paris Stock Exchange), the College of Surgery and the Treasury. In the dying days of the *Ancien Régime,* from 1787 onwards, Paris was encircled by a city wall with tollgates, which ran for 37km and effectively tripled the area covered by what was known as the 'greater inner city'. English, German and Russian travellers on the 'Grand Tour' now added Paris to their itinerary once more.

Rural exodus

Meanwhile, the economy of the city was booming. Confidence in the stability of the regime stimulated the growth of business and commerce, drawing hordes of people from the French provinces to come and seek their fortune in the capital and send back money to support their relatives. This wave of new workers found jobs as casual labour in the building trades, or worked in domestic service, textile mills or chemical plants. Certain regions specialised in different trades: the Auvergne, for instance, supplied water-sellers, wood-sellers, charcoal-sellers and boilermakers, stonemasons came from the Limousin region, chimney-sweeps from Savoy. These incomers, who were often illiterate, generally slaved away for more than ten hours a day and lived in miserable, cramped lodgings in the teeming working-class suburbs in central and eastern Paris. The west, around the Tuileries, Champs-Elysées and the Monceau plain, was home to the city's middle classes.

In 1781 the street now called the rue d'Odéon became the site of the first pavement in Paris. It was intended to keep pedestrians safe from carriages as they made their way to the nearby theatre in Luxembourg Gardens. The city prefect, Nicolas Frochot, tried hard to expand the pavement network, but the old city streets had many narrow archways, where footways were forced to stop short.

Public health and public entertainment

Gradually, gas street-lighting was installed; a census of 1810 listed 10,000 gas burners and 4,500 lamposts in the city. On the other hand, the streets were far from fragrant. In 1780, Louis XVI had banned the emptying of chamber pots into the street, but to no avail. Until the advent of flushing toilets in the late 19th century, the streets of Paris were awash with human and animal effluent. Rainwater and waste water from households ran off straight into the River Seine. Conditions inside houses were just as insanitary. Napoleon was renowned for his personal hygiene, taking frequent baths, dousing himself in Eau de Cologne and brushing his teeth, but most

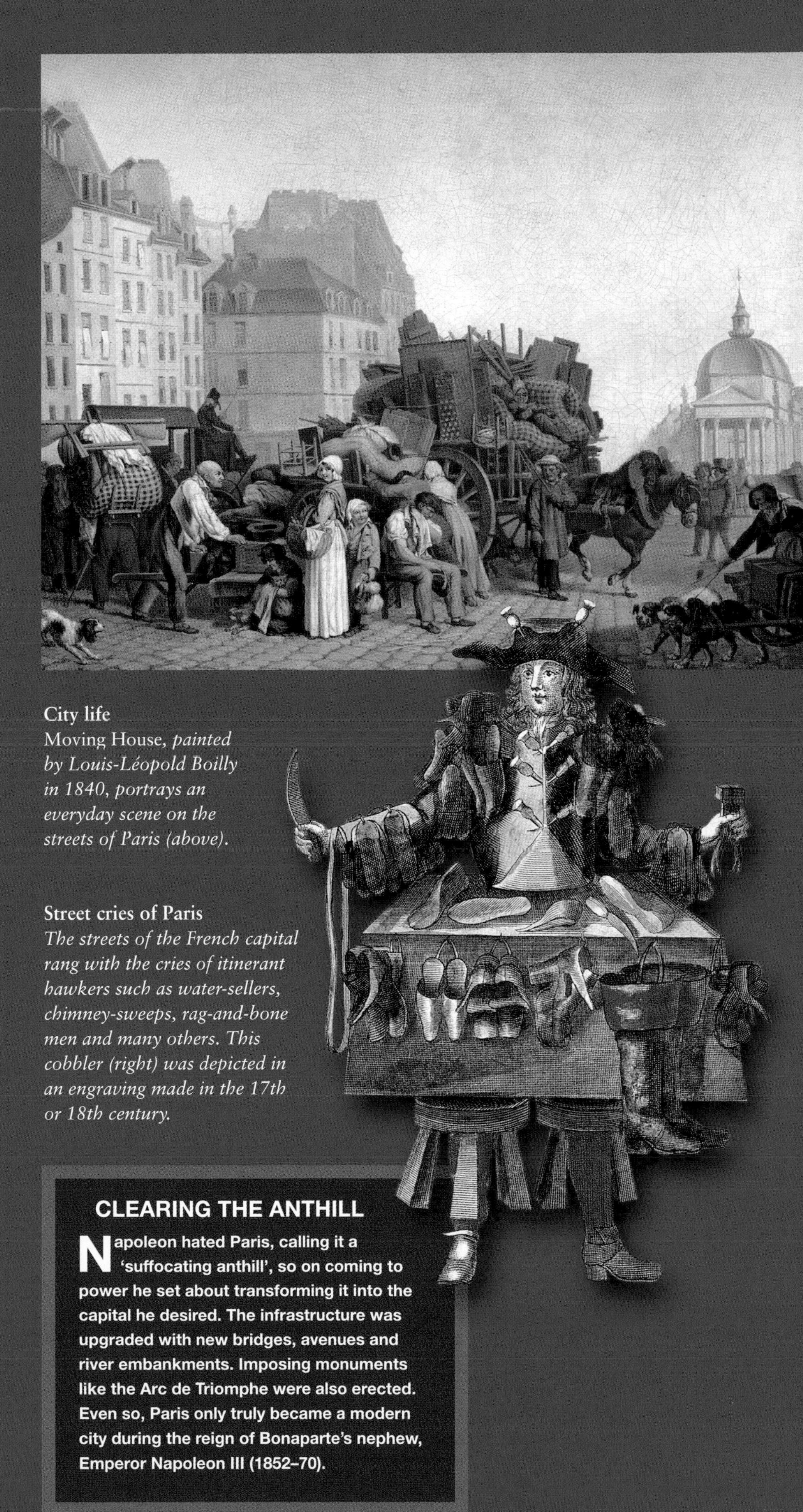

City life
Moving House, *painted by Louis-Léopold Boilly in 1840, portrays an everyday scene on the streets of Paris (above).*

Street cries of Paris
The streets of the French capital rang with the cries of itinerant hawkers such as water-sellers, chimney-sweeps, rag-and-bone men and many others. This cobbler (right) was depicted in an engraving made in the 17th or 18th century.

CLEARING THE ANTHILL

Napoleon hated Paris, calling it a 'suffocating anthill', so on coming to power he set about transforming it into the capital he desired. The infrastructure was upgraded with new bridges, avenues and river embankments. Imposing monuments like the Arc de Triomphe were also erected. Even so, Paris only truly became a modern city during the reign of Bonaparte's nephew, Emperor Napoleon III (1852–70).

Café society
One of the many new cafés to open in Paris was the fashionable Terrasse de la Rotonde, founded in 1810 in Montparnasse, shown here in an 1814 lithograph by George Emanuel Opitz.

Elegant living
A teacup and saucer decorated with gilded vine leaves and grapes, manufactured between 1787 and 1820 in the Locré-Russinger porcelain works, one of the foremost makers of bone china in Paris.

FINE DINING

The 'Provençal Brothers', one of the most popular restaurants in Napoleonic Paris, introduced Parisians to southern French food such as bouillabaisse and salt cod. Another top dining spot was the 'Great London Tavern' on the rue de Richelieu. Its proprietor, Antoine Beauvilliers, had formerly cooked for royalty and served his clientele in full regalia, with his sword at his side, in sumptuous surroundings of mahogany tables, crystal chandeliers and damask tablecloths.

Parisians were strangers to bathing. To be fair, the prices demanded by water-sellers (2 sous for 15 litres) was beyond most people's means.

The insanitary conditions did not stop Parisians from making merry in cafés, which mushroomed during this period – there were over 4,000 by 1807 – or from flocking to the new 'restaurants'. The Revolution marked a turning point in French eating habits. The great chefs of the *Ancien Régime*, who had once served the nobility, set up eating houses catering to the general public. The number of restaurant tables in Paris doubled from 500 in 1805 to 1,000 by 1825, and the city became the gastronomic capital of Europe. The new social élite consulted guides such as the *Gourmands' Almanac* by the food expert Grimod de la Reynière. 'Russian' service – with dishes coming one after the other, served in individual portions – supplanted traditional French service, where all the dishes came at the same time and diners served themselves.

Home comforts and the arts

Ordinary people's homes became more stylish and comfortable, as mass-produced wallpaper replaced limewashed walls. This new, relatively inexpensive decorative wall covering enabled people to imitate the lavish interior designs of the rich. Apartments had once been heated by a single hearth with a mantelpiece generally made from oak, sometimes from beech, hornbeam or lime. New buildings boasted several fireplaces in a wide variety of materials. Before retiring to bed, people stoked up their fires with logs floated down the Seine, or with more expensive fresh-cut wood brought to the capital by boat. As the winter temperature outside plummeted, the inhabitants of the new Paris were warm and snug in their beds. In well-off families, everyone had their own bed, although the poor still shared.

Sculpture enjoyed a revival. Napoleon loved to visit the Louvre at night to view the ancient Greek, Roman and Egyptian statues by torchlight. In painting, Neoclassicism was the predominant style and the leading exponent was Jacques-Louis David (1748–1825), who found himself swamped with commissions. Music lovers were catered for by the 'Concert des Amateurs', which opened on the rue de Cléry in 1799. Aficionados regarded it as being among the best venues in Europe for symphonic performances. Soon after, in 1801, two opera companies performing the Italian-language works of Mozart set up on the banks of the Seine. A census of 1803 listed no fewer than 17 theatres in Paris. As one contemporary commentator wrote: 'There's a veritable epidemic of theatres; it seems that there is no run-down church or covered space of any reasonable size that isn't being taken over and converted into a playhouse.'

Seat of learning

Under Napoleon, Paris developed into a major centre of learning attracting scholars from around Europe. The Academy of Sciences, briefly abolished by the Jacobins in 1793, was replaced by the Society of Physical and Mathematical Sciences, comprising 66 leading scholars of the age. The Polytechnic College – founded in September 1794 and housed temporarily in outbuildings of the former Bourbon royal palace – recruited its teaching staff from among the nation's foremost scientists. The institution became the standard route for those who wanted to pursue a scientific or technical career.

Anyone wishing to stay abreast of the latest scientific developments read the *Philosophical, Literary and Political Journal*. Founded in the dark days of the Terror, the publication devoted a third of its space to scientific subjects and ran until 1807; it appeared every 10 days in accordance with the 10-day week of Revolutionary France. Finally, to help Parisians get used to the new metric system, the authorities placed marble measurement gauges at 16 busy locations, including the rue de Vaugirard and the Place Vendôme.

CITY OF SCIENTISTS

In the final days of Louis XVI, all of France's leading scientists were resident in Paris. In 1787 the mathematician Joseph Louis Lagrange (originally Giuseppe Lodovico Lagrangia from Turin), gave up his post at the Academy in Berlin and moved to Paris to join his eminent colleagues Laplace, Condorcet, Monge, Delambre and Lavoisier. At that time, no other capital city could boast such an impressive array of research institutions, including the Royal Academy of Science, the King's Botanical Gardens and Natural History Collection, and the Paris Observatory.

Imperial symbol
One of the emblems used by Napoleon was a hive with bees, which symbolised immortality. The emblem originally came from the tomb of the 5th-century Frankish king, Childeric I. This plaster frieze of a hive with bees is on the balcony of the Comédie Française.

Seine crossing
When it was begun in 1787, the Pont de la Concorde (below) was known as the Pont Louis XVI, then became the Pont de la Révolution. It was finished in 1791, in part with blocks of stone from the Bastille prison, which was stormed and torn down in 1789.

CHRONOLOGY

The timeline on the following pages outlines the key discoveries and inventions around the early years of the Industrial Revolution. Selected historical landmarks are included to provide chronological context for the scientific, technological and other innovations listed in the columns below them.

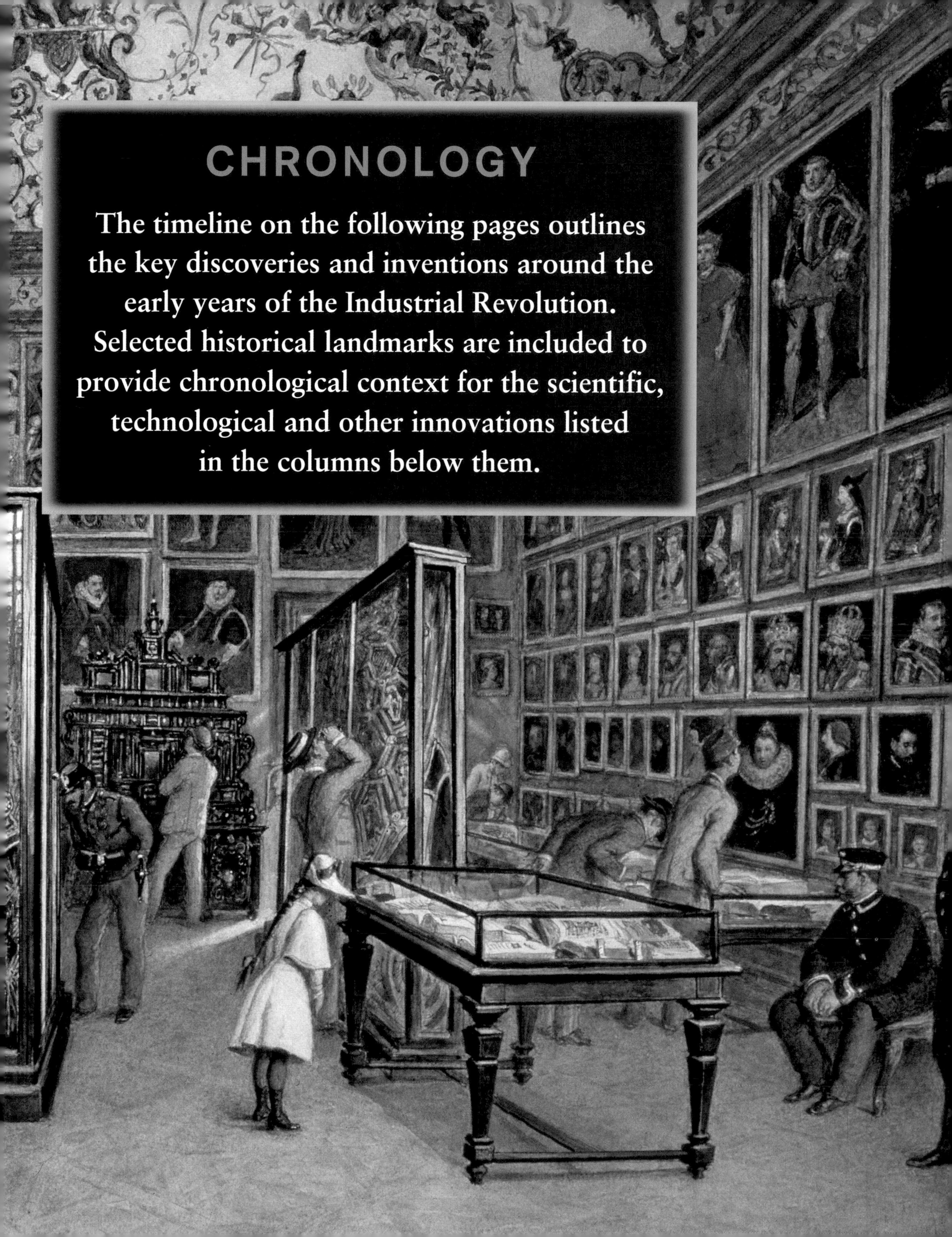

1745

EVENTS

- War of the Austrian Succession (1740–48)
- Frederick II of Prussia builds Sans Souci palace in Potsdam (1745–47)
- Defeat of the Jacobite uprising in Scotland and destruction of the clan system (1745–46)

INVENTIONS

- Jacques de Vaucanson, known as a maker of automata, develops an automatic tape-controlled loom for weaving silk
- John Roebuck introduces the lead-chamber method for the industrial manufacture of sulphuric acid in Birmingham
- The Leyden jar, the first practical way to store static electricity, is invented almost simultaneously by Pieter van Musschenbroek and Ewald Georg von Kleist
- German chemist Andreas Sigismund Marggraf discovers that sugar can be derived from sugar beet
- French scholar François Fresneau writes the first scientific paper on the potential uses of the rubber tree *(Hevea brasiliensis)*

1750

EVENTS

- Death of the composer Johann Sebastian Bach (1750)
- The Chinese Qing dynasty gains control over Tibet (1751)
- The unification of Burma begins under King Alaungpaya (1752)

INVENTIONS

- Diderot and d'Alembert publish the first volume of the *Encyclopédie*, a groundbreaking new reference work that becomes the quintessential text of the Enlightenment
- Swedish chemist Baron Axel Fredrik Cronstedt discovers nickel
- Nicolas Louis de la Caille and Joseph Jérôme Lefrançois calculate the distance from the Earth to the Moon
- American statesman and scientist Benjamin Franklin demonstrates that lightning is an electrical discharge and invents the lightning rod
- English chemist Joseph Black discovers carbon monoxide

▼ A gathering of Enlightenment philosophers and editors of the *Encyclopédie*

▼ French scientist Thomas-François Dalibard conducting an experiment with a lightning rod

▼ Portrait of Jean le Rond d'Alembert in the frontispiece to the first volume of the *Encyclopédie*

1755

- Lisbon is destroyed by an earthquake (1755)
- The Seven Years' War ravages Europe (1756–63)
- British forces begin the conquest of India (1757–1849)

- Swiss naturalist Albrecht von Haller outlines the basic principles of physiology, in particular the role of the brain and nervous system in the movement of muscles
- The British Museum in London opens as the world's first public museum

▲ A bust of the Emperor Constantine displayed in the Capitoline Museum, Rome, one of the world's first public museums

1760

- Jean-Jacques Rousseau publishes his influential work *The Social Contract* (1762)
- The Peace of Paris grants Britain all French possessions in Canada and west of the Mississippi (1763)
- Jesuits banned in France, Spain, Naples, Bavaria and Venice (1764)

- English mapmaker John Spilsbury creates the jigsaw puzzle
- The world's first roller skates are made by a Belgian watchmaker Jean-Joseph Merlin
- Giovanni-Battista Morgagni establishes the discipline of modern anatomical pathology
- The sandwich is invented and named after the avid gambler John Montagu, Earl of Sandwich
- First use of tobacco juice as a pesticide against aphid infestation in France

▲ Early 20th-century advertising poster promoting the health benefits of roller skating

◄ A portrait of John Montagu, Earl of Sandwich, by Joseph Highmore

1765

EVENTS

- French scientific expedition led by Louis-Antoine de Bougainville sets sail to circumnavigate the globe (1766)
- First Pacific voyage of British maritime explorer Captain James Cook (1768)

INVENTIONS

- British chemist Henry Cavendish discovers hydrogen, a gas 14 times lighter than air
- Richard Reynolds installs cast-iron rails in the coal mines at Coalbrookdale, Shropshire, and fits the coal wagons that run on them with brake shoes, operated by a hand lever
- The first modern circus, featuring mainly displays of horsemanship, is founded in London by former cavalry officer Philip Astley
- Publication of the final volume of Diderot and d'Alembert's *Encyclopédie*

1770

EVENTS

- James Cook sails into Botany Bay (1770)
- First Partition of Poland by Prussia, Russia, and Austria (1772)
- American colonists stage the Boston Tea Party in protest at taxes on tea imposed by the British government (1773)
- Technical University founded in Istanbul (1773)

INVENTIONS

- Joseph Priestley, a chemist specialising in the study of gases, succeeds in isolating oxygen
- Richard Arkwright founds the modern factory system based on shift work at his cotton-spinning mill at Cromford, Derbyshire

▼ Antoine Laurent de Lavoisier and his wife, in a portrait by French Neoclassical painter Jacques-Louis David

▲ Model of a railway wagon incorporating wooden brake shoes

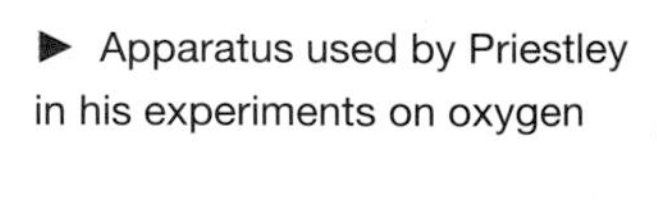

► Apparatus used by Priestley in his experiments on oxygen

1775

- Captain James Cook is killed during his third voyage of exploration to the Pacific (1776–82)
- American Declaration of Independence from Britain (1776)
- The Spanish found the Viceroyalty of Río de la Plata in what will later become Argentina (1776)

- French scholar Antoine Laurent de Lavoisier establishes the foundations of modern chemistry; encapsulated in his Law of the Conservation of Mass: 'Nothing is created, nothing is lost, everything is transformed'
- The world's first iron bridge is constructed at Coalbrookdale

1780

- The Fourth Anglo-Dutch War (1780–84)
- Rama I, King of Siam, founds the royal dynasty that continues to rule Thailand to the present day (1782)
- Russia annexes the Crimea (1783)

- Italian scientist Lazzaro Spallanzani performs the world's first artificial insemination – on a dog – thereby confirming the role of spermatozoa in sexual reproduction
- First meteorological society is founded in Germany by Karl Theodor, duke of Bavaria
- French watchmaker Abraham Louis Breguet invents the button watch winder
- Using telescopes built to his own specifications, English astronomer William Herschel discovers Uranus
- Arnould Carangeot devises the goniometer, an instrument for measuring angles that is subsequently adapted for many different uses
- Inspired by Priestley's discovery of lighter-than-air gases, the brothers Étienne and Joseph Montgolfier construct the world's first hot air balloon
- French inventor François de Jouffroy d'Abbans launches the first steam-powered boat on the Saône River in Lyons

► Abraham Louis Breguet's No.15 self-winding watch, made in 1780

▲ Trial run of the steamboat *Pyroscaphe* on the Saône River in France

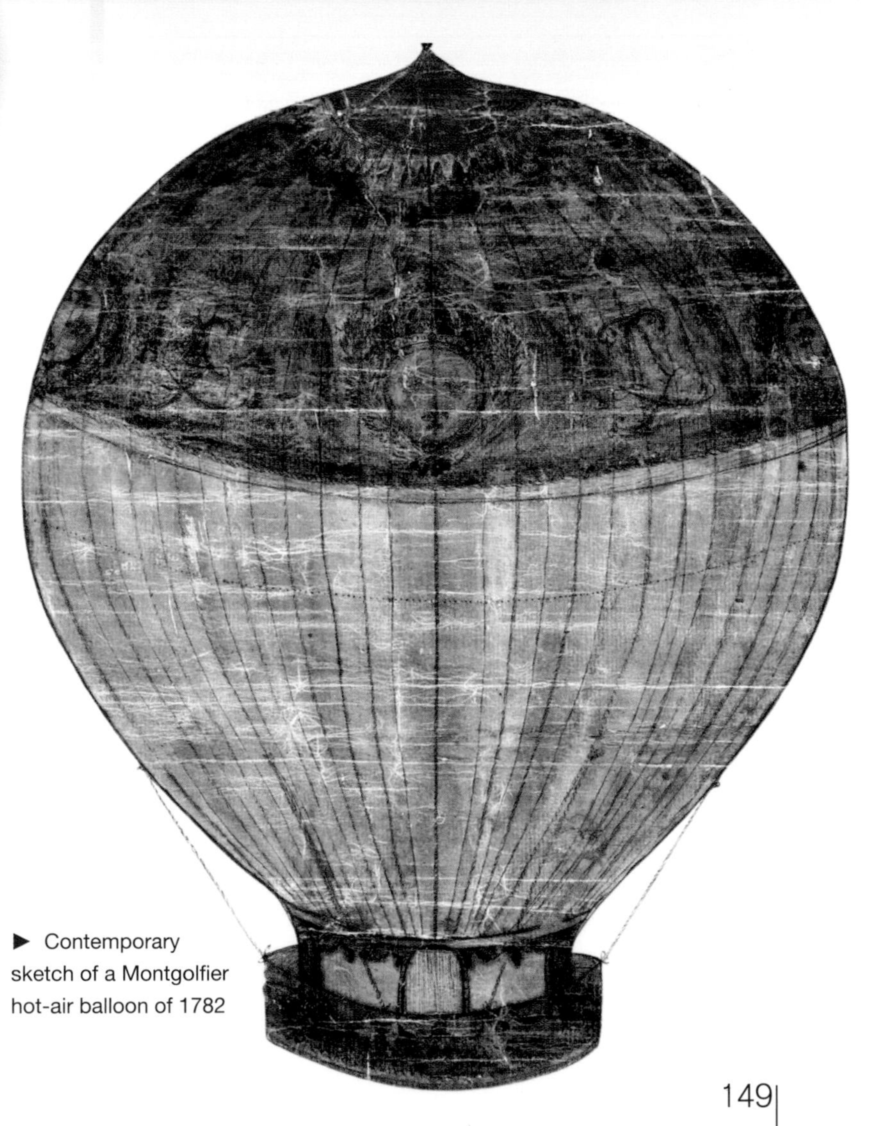

► Contemporary sketch of a Montgolfier hot-air balloon of 1782

1785

EVENTS

- Chronic rice shortage leads to famine and riots in Japan (1787)
- Outbreak of the French Revolution and the *Declaration of the Rights of Man and of the Citizen* (1789)
- George Washington is inaugurated as first president of the United States of America (1789–97)

INVENTIONS

- French scientific expedition commanded by the Comte de La Pérouse vanishes in the Pacific
- Edmund Cartwright patents the first power loom
- Balloonist Jean-Pierre Blanchard invents the parachute
- Parisian dentist Nicolas Dubois de Chémant submits a patent for porcelain false teeth
- British shipwright Henry Greathead constructs the first purpose-built lifeboat

1790

EVENTS

- The southern United States depends on the forced labour of an estimated half a million slaves (*c*1790)
- Mozart dies after composing the opera *The Magic Flute* and while working on an unfinished requiem mass (1791)
- Louis XVI makes an unsuccessful attempt to flee Revolutionary France (1791); he is beheaded two years later

INVENTIONS

- Highwaymen Nicolas Pelletier becomes the first person to be executed by guillotine
- Astronomers Jean-Baptiste Delambre and Pierre Méchain measure part of the Earth's meridian between Dunkirk and Barcelona, thus laying the foundations of the metric system of measurement
- The optical telegraph, with a pair of movable arms to signal letters of the alphabet, is invented by French physicist Claude Chappe
- Scottish engineer William Murdock harnessed coal gas for lighting
- In America, cotton spinning enters the industrial age with the introduction of Eli Whitney's cotton gin

▶ A Maurand disc calculator

▲ Replica of Eli Whitney's cotton gin

◀ Henry Greathead's *Original*, the world's first true lifeboat

1795

- British explorer Mungo Park leads an expedition to Africa (1795–97)
- Death of Catherine II, the Great, of Russia (1796)
- Napoleon Bonaparte overthrows the French government in a coup and replaces it with the Consulate (1799-1804)
- Nelson defeats the French at the Battle of the Nile (1798)

- English clergyman Samuel Henshall invents the modern corkscrew
- French chef Nicolas Appert develops a method of preserving food by sealing it hermetically
- German physician Christian Hahnemann evolves a controversial new therapy known as homeopathy
- British doctor Edward Jenner paves the way for the fight against infectious diseases with the first inoculation against smallpox
- German actor and playwright Alois Senefelder invents lithography
- British engineer Henry Maudslay constructs the first machine tool: a slide-rest lathe, ancestor of all modern metal lathes

1800

- The United Kingdom is formed by the union of Great Britain and Ireland (1800)
- Founding of the Swiss Confederation (1803)
- Thomas Jefferson purchases Louisiana from France (1803)
- Civil Code enacted in France (1804)

- Joseph Marie Jacquard's card-operated automatic loom streamlines textile manufacture
- British astronomer William Herschel identifies the existence of an invisible form of light – infrared radiation
- Johann Wilhelm Ritter, a German chemist, discovers ultraviolet light
- Italian physicist Alessandro Volta introduces the electric pile, capable of supplying a constant current
- Invention of the chromatic trumpet, which can play all the notes of the scale
- British doctor John Dalton lays the foundation of modern atomic theory

▼ Page from a 19th-century catalogue advertising a selection of corkscrews

▲ A Jacquard loom

► A Senot screw-cutting lathe of 1795

1805

EVENTS

- Nelson destroys French and Spanish fleet at Trafalgar (1805)
- Mehmet Ali become Viceroy of Egypt (1805)
- Dissolution of the Holy Roman Empire by Napoleon (1806)
- Abolition of the slave trade in the British Empire (1807)
- Start of the Spanish War of Independence (1808–14)

INVENTIONS

- Friedrich Sertürner isolates morphine, the active narcotic ingredient of opium
- Italian chemist Luigi Brugnatelli invents electroplating, a method of depositing a thin layer of metal on objects electrochemically
- Carbon paper is invented by Ralph Wedgwood
- English inventor William Congreve creates the first modern rockets, which are deployed as weapons of war

1810

EVENTS

- Napoleon's Russian campaign ends in defeat at the Battle of Berezina (1812)
- Mexico proclaims independence from Spain (1813)
- End of the 'War of 1812' between Britain and the USA (1814)
- Napoleon is deposed and exiled to Elba (1814)

INVENTIONS

- Joseph von Fraunhofer invents the spectroscope, which enables scientists to examine the spectra of distant stars and analyse their chemical composition
- André-Marie Ampère is the first to recognise that molecules are composed of atoms
- Swedish chemist Jöns Jacob Berzelius pioneers the use of chemical symbols

▼ Alexandra, Princess of Wales, being shown the electroplating process

► A Congreve rocket

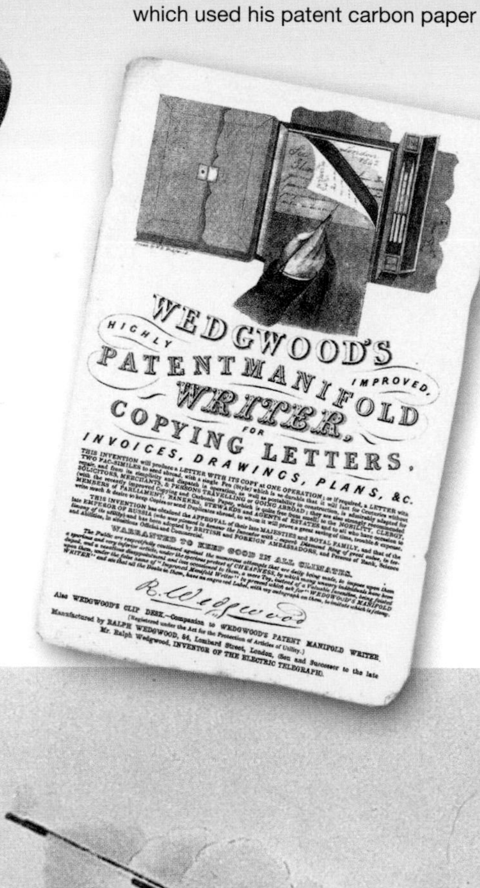

▼ An advertisement for Wedgwood's copying machine, which used his patent carbon paper

▲ Infantry unit firing rockets, from a work by William Congreve

1815

• The Congress of Vienna redraws the borders of Europe following Napoleon's exile to Elba (1814–15)
• Final defeat of Napoleon at the Battle of Waterloo and restoration of the French Bourbon monarchy (1815)

• René Laennac invents the stethoscope

• Humphrey Davy's safety lamp replaces oil lamps in mines, which were extremely dangerous in the presence of firedamp (methane gas)

• Joseph Maelzel introduces the metronome

• Scottish physicist Sir David Brewster invents the kaleidoscope

• French chemist Louis Jacques Thénard discovers hydrogen peroxide, which finds a ready application as a pulp bleaching agent in the paper manufacturing industry

• John Loudon MacAdam develops the road paving method named after him

• Patents are submitted by a number of inventors for hand-held small arms or pistols

• The diving helmet is introduced by the German Augustus Siebe

1820

• Wars of Independence are fought throughout Latin America (1815–25)
• Founding of the colony of Monrovia (modern Liberia) on the west coast of Africa as a home for freed slaves from the USA (1821)
• Brazil gains independence from Portugal (1822)

• Czech physiologist Jan Evangelista Purkinje proposes a system for classifying fingerprints, later used in criminology

• Charles Mackintosh invents the rubberised waterproof fabric that bears his name

• English mathematician Charles Babbage develops the 'Difference Engine', a machine for calculating logarithms and trigonometric functions

• French physicist Nicolas Léonard Sadi Carnot proposes the theory of the Carnot cycle, the basis of the science of thermodynamics

► The Rosetta Stone, which provided the key to the deciphering of ancient Egyptian hieroglyphs

► Terrace of the café La Rotonde in Paris in the early 19th century

Index

Page numbers in *italics* refer to captions.

T

U

V

W

YZ

Picture credits

ABBREVIATIONS : t = top, c = centre, b = bottom, l = left, r = right

Front cover: main image, trial run of an early steamboat on the River Saone in France, Leonard de Selva/Corbis DS001494; inset: a Breguet watch, © Montres Breguet SA.
Spine: early weaving shuttles, The Bridgeman Art Library/Science Museum, London.
Back cover: Eli Whitney's cotton gin, Kharbine-Tapabor Collection.
Page 2, left to right, top row: Rue des Archives/PVDE; Cosmos/SSPL/Science Museum; Cosmos/SSPL/Science Museum; 2nd row: SRD/Jean-Pierre Delagarde/with the kind permission of the Electronic Laboratory – ESPCI, Paris; Leemage/Costa; Cosmos/SSPL/Science Museum; 3rd row: The Bridgeman Art Library/Archives Charmet, Musée de la Poste, Paris; Corbis/Bettmann; Cosmos/ SSPL/Science Museum; bottom row: Cosmos/SSPL/Science Museum; Cosmos/SSPL/Science Museum; © Paris – Musée de l'Armée, Dist/RMN/É Cambier.
Pages 4/5: the Iron Bridge at Coalbrookdale, © Mike Hayward, photoshropshire.com; 6t: Corbis/Bettmann; 6b: Hemis FR/C Moirenc; 6/7c & 7bl: AKG-Images/ Bibliothèque Louis-Aragon, Amiens; 7tl: Cosmos/SSPL/Science Museum; 7tr: Horizon Features/M Viard; 7br: Leemage/Selva private collection; 8tl: Cosmos/SSPL/Martin Bond, Leeds; 8b: AKG-Images; 8/9: Corbis/Historical Picture Archive, 9t & bl: Cosmos/SSPL/Science Museum; 9br: Corbis/R Estall; 10tl: Leemage/AISA University Library, Barcelona; 10bl: Leemage/Costa; 10/11: The Bridgeman Art Library/private collection; 11tl: The Bridgeman Art Library/Giraudon/Musée Carnavalet, Paris; 11tr: Cosmos/SSPL/Science Museum; 11b: Kharbine-Tapabor/Collection Jean Vigne; 12/13: Kharbine-Tapabor Collection; 12bl: Leemage/Photo Josse, Conservatoire national des Arts et Métiers, Paris; 12br & 13t: Cosmos/SSPL/ Science Museum; 13cr: AKG-Images/L Dodd, 1943 © Adagp, Paris 2009; 13b: The Bridgeman Art Library/The Stapleton Collection; 14tl: AKG-Images/A Gerle, Sotheby's; 14bl: RMN/DR, L-E Barrias, Musée de la Musique, Paris; 14/15 & 15tl: Cosmos/SSPL/Science Museum; 15c: Leemage/Heritage Image/Ann Ronan Picture Library; 17r: Leemage/AISA Deutsches Museum, Munich; 16tl: Cosmos/SSPL/Science Museum; 16c: Leemage/Angelo, L Cogniet, Musée du Louvre, Paris; 16r: © Paris – Musée de l'Armée, Dist RMN/É Cambier; 17t: National Army Museum, London; 17c: RMN/DR, Musée d'Orsay, Paris; 17b: Eyedea/Hoa-Qui/P Escudero; 18/19: Corbis/S Bianchetti; 20: AKG-Images/Erich Lessing/Musée de la Coopération Franco-Américaine, Blérancourt; 21r: Leemage/Selva private collection; 21l: Leemage/Selva private collection; 21b: Corbis/ Weatherstock; 22t: Corbis/Bettmann; 22b: The Bridgeman Art Library/Giraudon/Musée d'Art et d'Histoire, Saint-Germain-en-Laye; 23l: Rue des Archives/PVDE; 23r: RMN/H Lewandowski, Photo G Loppé, Musée d'Orsay, Paris; 24: Leemage/ Selva private collection; 25t: AKG-Images/ E Lessing, Bibliothèque Nationale, Paris; 25b: Leemage/Photo Josse, private collection; 26l: The Bridgeman Art Library/Giraudon/Lauros/private collection; 26r: La Collection/Domingie & Rabatti, private collection, Florence; 27t: AKG-Images; 27bl: The Bridgeman Art Library, Jules Michelet, T Couture, Musée de la Ville de Paris, Musée Carnavalet, Paris; 27br: The Bridgeman Art Library/The Stapleton Collection; 28: Corbis/T Pannell; 29t: Corbis/Sygma/B Annebicque; 33tr: Leemage/Heritage Image/W Radclyffe; 30c: The Picture Desk/Coll Dagli Orti/Gianni Dagli Orti, Capitoline Museum, Rome; 31t : AKG-Images/ Kunshistorisches Museum, Vienna; 31b: Hemis FR/C Moirenc; 32tl: Eyedea/Keystone/M Evans; 32b: Cosmos/SSPL/Science Museum; 33t: The Bridgeman Art Library/Barbara Singer/Private Collection; 33b: Corbis/Zefa/Roy McMahon; 34l: Rue des Archives/The Granger Collection, New York, J Highmore; 34tr: The Bridgeman Art Library/Dato Images/Private Collection, T Etbauer, DR; 34b: Corbis/Photocuisine/Turiot-Roulier; 35t: AKG-Images/Bibliothèque Louis Aragon, Amiens; 35b: Corbis/V D Schreck; 36: Horizon Features/ M Viard; 37t: Corbis/G Rowell; 37b: Corbis/EPA/ P Pleul; 38t: The Picture Desk/The Art Archive/ E Tweedy; 38b: Cosmos/SSPL/Martin Bond, Leeds; 39: Rue des Archives/Epimedia; 40t: Cosmos/SSPL; 40c: Cosmos/SSPL/Phantamix; 40b: Cosmos/ SSPL/Dr Keith Wheeler; 41t & r: Cosmos/SSPL; 44t: © Musée des Arts et Métiers-Cnam, Paris/ Photo C Le Toquin; 42b: AKG-Images; 43: The Picture Desk/The Art Archive/Culver Pictures; 44tl: Corbis/Swim Ink 2, LLC; 44tr: Corbis/Historical Picture Archive; 44b: Eeyedea/Rapho/H Bruhat; 45: The Bridgeman Art Library/Archives Charmet Académie des Sciences, Paris; 46t: Cosmos/ SSPL/Science Museum; 46b: The Bridgeman Art Library/Giraudon/Lauros/Archives Charmet, Conservatoire national des Arts et Métiers, Paris; 46/47: The Bridgeman Art Library/Archives Charmet, Bibliothèque des Arts décoratifs, Paris; 47b: Cosmos/SSPL/ESA; 48t: Leemage/Photo Josse, Conservatoire national des Arts et Métiers, Paris; 48b: Lycée Louis-le-Grand, Musée Pierre-Provost, Paris; 49: Cosmos/SSPL/Mehau Kulyk; 50t: Rue des Archives/CCI; 50b: Leemage/Selva private collection; 51t: RMN/Hervé Lewandowski, A-J Dalou, Musée d'Orsay, Paris; 51b: AKG-Images/E Lessing, J-L David, Metropolitan Museum of Art, New York; 52tl: Corbis/R Estall; 52b: Eyedea/Hoa-Qui/P Escudero; 53tr: Corbis/Bettmann; 53c: Cosmos/SSPL/Science Museum; 54t, 54b, 54b: Cosmos/SSPL/Science Museum; 55t: Cosmos/SSPL/NASA; 56t: Leemage/AISA, University Library, Barcelona; 56b: Leemage/AISA, private collection, Barcelona; 57t: The Bridgeman Art Library/F-S Ravenet © with the kind permission of the Oxford Colleges; 57b: Cosmos/SSPL/Juergen Berger; 58t: © Montres Breguet SA; 58c: SRD/Jean-Pierre Delagarde/with the kind permission of the Electronic Laboratory – ESPCI, Paris; 59t: Cosmos/SSPL/NASA/ESA/STSCI/E Karkoschka, University of Arizona; 59b, 60t: Leemage/Costa; 60b: AKG-Images/Lourenco de Gusmaos Passarola, 1709, 1937, O Nerlinger, DR; 61t: The Bridgeman Art Library/private collection; 61b: The Bridgeman Art Library/Archives Charmet, Musée Carnavalet, Paris; 62t: Leemage/Photo Josse, Bibliothèque nationale, Paris; 62/65: Corbis/S Bianchetti; 63t: Corbis/Bettmann; 64: Leemage/Photo Josse, J Didier, J Guiaud, Musée Carnavalet, Paris ; 65t: AKG-Images; 65cl: Corbis/Bettmann; 65b: Corbis/C J Morris; 66t: Corbis/Bettmann; 66b: Corbis/Léonard de Selva; 67b: Corbis/Bettmann; 67t, 68bl, 68/69: Cosmos/SSPL/Science Museum; 70t: Corbis/Robert Harding World Imagery/Bruno Barbier; 70c: Leemage/Selva private collection; 70b: Kharbine-Tapabor/Collection Perrin, Worms & Cie, illustration Sandy Hook, DR; 71c: Cosmos/ SSPL/Science Museum; 71b: The Bridgeman Art Library/Look and Learn/Private collection, G Coton; 72t: Cosmos/SSPL/Science Museum; 72c: Leemage/Selva private collection; 72b: AKG-Images/E Lessing, Roelof Koets the Elder, Museum Mayer van der Bergh, Anvers; 73t: The Bridgeman Art Library/Giraudon/Musée Carnavalet, Paris; 73b: AFP; 74: The Bridgeman Art Library/Giraudon, 15th-century fresco, Castello di Issogne, Val d'Aoste; 75t: Kharbine-Tapabor/Collection Jonas, illustration by Daniel Vierge; 75c: Leemage/Photo Josse, Jean-Baptiste Joseph Delambre, 1879, Henri Coroenne, 1822-1909, Musée de l'Observatoire, Paris; 75br: AKG-Images/ Bibliothèque nationale, Paris; 76t: Kharbine-Tapabor/Collection Jean Vigne; 76b: Cosmos/ SSPL; 77l : Cosmos/SSPL/Science Museum; 77r: © Musée des Arts et Métiers-Cnam, Paris/Photo P Faligot; 78t: Eyedea/Keystone/ M Evans; 78b: Cosmos/SSPL; 79t: Corbis/ Tempsport/Olivier Prevosto; 79b: Eyedea/Hoa-Qui/AGE/E Coelfen; 80t: Leemage/LEE, A Bitard; 80b: The Picture Desk;The Art Archive/Gianni Dagli Orti, Bibliothèque des Arts décoratifs, Paris; 81t: AKG-Images/A Gerle, Sotheby's; 81b: Eyedea/ Keystone France; 82: The Bridgeman Art Library/ Archives Charmet, Musée de la Poste, Paris; 83l: Leemage/Selva private collection; 83c: © Coll. Musée de la Poste, Paris/La Poste; 83r: Leemage/Photo Josse Conservatoire national des Arts et Métiers, Paris; 84t: Leemage/Photo Josse Conservatoire national des Arts et Métiers, Paris; 84b: Cosmos/SSPL/S Terry; 85t: Corbis/ Bettmann; 85b: The Bridgeman Art Library/The Stapleton Collection; 86t: Roger-Viollet; 86b: Cosmos/SSPL/Science Museum; 87: Kharbine-Tapabor Collection; 88t: AKG-Image/Coll. Archiv für Kunst & Geschichte; 88bl: Corbis/Photocuisine/Viel; 88br: Illustration (see below); 89t: Cosmos/ SSPL/Science Museum; 89b: AKG-Images/ A Warhol, José Mugrabi collection, New York © The Andy Warhol Foundation for the Visual Arts, New York/Adagp, Paris 2009; 90tr: Hemis FR/AGE/ Peter Holmes; 90bl: Cosmos/SSPL/Science Museum; 91t: The Bridgeman Art Library/Archives Charmet, private collection; 91b: The Bridgeman Art Library/Lady Mary Wortley Montagu, J Richardson, Galleries and Museums Trust © Sheffield Museums; 92tl: Leemage/Photo Josse/G Melingue, Académie de médecine, Paris; 92c: Cosmos/ SSPL/Science Museum; 93t: Corbis/EFE/L Alonso; 93b: Cosmos/SSPL/Eye of Science; 94tl: © National Maritime Museum, Greenwich, London; 94bl: Leemage/Photo Josse, A-L Garneray, Manufacture de porcelaine, Sèvres; 95t: AKG-Images/L Dodd, 1943 © Adagp, Paris, 2009; 95b: The Bridgeman Art Library/Archives Charmet, private collection; 96t: Hemis FR/C Heeb; 96b: AKG-Images/Ullstein Bild; 97tl: Leemage/ Heritage Images, British Library, London; 97: AKG-Images/Coll. Archiv für Kunst & Geschichte; 98tr: The Bridgeman Art Library/ Archives Charmet, Bibliothèque Mazarine, Paris; 98b: AKG-Images/E Lessing, N Monsiau, Château et Trianons, Versailles; 99t: AKG-Images/ F G Weitsch, Staatliche Schlösser und Gärten, Berlin; 99b: © National Maritime Museum, Greenwich, London; 100c: Leemage/Bianchetti; 100bl: Cosmos/SSPL/Science Museum; 101t: The Bridgeman Art Library/The Stapleton Collection, F-L Gottlob, private collection; 101b: AKG-Images/Sarah Bernhardt, 1896, A Mucha, Impression F Champenois, Victor Arwas collection © Mucha Trust/Adagp, Paris 2009; 102t: Leemage/Photo Josse, private collection; 102b: © Musée des Arts et Métiers-Cnam, Paris/photo Studio Cnam; 103t: © Musée des Arts et Métiers-Cnam, Paris/photo Studio Cnam; 103b, 104tl, 104br, 105t: Cosmos/SSPL/Science Museum; 105b: © BPK, Berlin Dist. RMN Photographie de Georg Buxenstein & Co c1900; 106t: The Picture Desk/The Art Archive/Gianni Dagli Orti, Écomusée de la Communauté Le Creusot; 106b: Eyedea/Hoa-Qui/M Troncy; 107t: AKG-Images/Fritz Lang; 107b: RÉA/ROPI/I Kazmierczak; 108t: AKG-Images/E Lessing, Haydn Museum, Rohrau; 108b: The Picture Desk/The Art Archive/ Gianni Dagli Orti, Museum der Stadt, Vienna; 109t: Collection Christophel, Ingmar Bergman; 109b: RMN/DR, L-E Barrias, Musée de la Musique, Paris; 110t: The Bridgeman Art Library/Ronzi, M Gauci © City of Westminster Archive Centre, London; 110b: AKG-Images/C Schloesser; 111t: AKG-Images/E Lessing Gesellschaft der Musikfreunde, Vienne; 111b: Corbis/Lebrecht Music & Arts; 112l: Leemage/Luisa Ricciarini, Archive of the Abbey of Monte Cassino; 112b: Andia FR/ Mattes; 113t: The Bridgeman Art Library/Science Museum, London; 113b: Cosmos/SSPL/Science Museum; 114t: © Musée des Arts et Métiers-Cnam, Paris/Photo C Le Toquin; 114b: © Musée des Tissus de Lyon/P Verrier; 115t: Leemage/Costa; 115b: Leemage/Heritage Images/D Burrows; 116tl: Cosmos/SSPL, G Tissandier; 116b: Corbis/ M Nicholson; 117t: fedephoto.com/O Coret; 117b: RÉA/R Damoret; 118l: fedephoto.com/ B Bakalian; 118tr: Leemage/Heritage Images/ Museum of London; 118ctr: REA/Panos/P Barker; 118cbr: The Picture Desk/The Art Archive/Alfredo Dagli Orti; 118b: REA/Laif/T Reinicke; 119tl: Andia FR/Alpaca/E Soudan; 119cl: Cosmos/SSPL/Science Museum; 119bl: REA/R Damoret; 119tr: The Picture Desk/The Art Archive/Alfredo Dagli Orti; 119cr: Corbis/Gianni Dagli Orti; 119br: Corbis/The Stapleton Collection; 120t: Cosmos/SSPL/Science Museum; 120b: Leemage/Heritage Images/Ann Ronan Picture Library; 121t: Cosmos/SSPL/Agema Infrared Systems; 121b: Cosmos/SSPL/T-Service; 122r: Leemage/Costa; 122b: Eyedea/Jacana/ Grandeur Nature/P Louisy; 123l: Leemage/AISA Deutsches Museum, Munich; 123b: AKG-Images, Bertini; 124t: Cosmos/SSPL/S Terry; 124b: Leemage/Selva private collection; 125t: Leemage/ Bianchetti; 125b: Eyedea/Keystone France; 126t: RMN/Matheus Châteaux de Malmaison et Bois-Préau, Malmaison; 126b: Leemage/Photo Josse, Musée de l'Observatoire, Paris; 127t: Leemage/Angelo, L Cogniet, Musée du Louvre, Paris; 127b: Leemage/AISA, British Museum, London; 128t: AKG-Images/G Mermet, D-V Denon, Paris, private collection; 128b: Leemage/Heritage Images; 129l: RMN/T Ollivier, Musée du Louvre, Paris; 129r: RMN/D Arnaudet, Musée du Louvre, Paris; 130tl: The Bridgeman Art Library/Archives Charmet, private collection; 130b: Cosmos/SSPL/ Science Museum; 131: Corbis/M Freeman; 132t: The Bridgeman Art Library © Newberry Library, Chicago; 132c: Cosmos/SSPL/Science Museum; 132b : BSIP/H Raguet; 133t: © Paris – Musée de l'Armée, Dist/RMN/É Cambier; 133b: Corbis/Mosaic Images; 134t, 134b, 135t: Cosmos/SSPL/Science Museum; 134c: RMN/DR, Musée d'Orsay, Paris; 135b: The Picture Desk/ The Art Archive/E Tweedy, private collection; 136/137: Leemage/Heritage Images/William Allen, British Library, London; 136b: Cosmos/SSPL/ Science Museum; 137c: The Bridgeman Art Library/The Stapleton Collection; 137r: The Bridgeman Art Library/The Stapleton Collection; 138t, 138b: National Army Museum, London; 139tr: Cosmos/SSPL/NASA; 139bl: Collection Christophel, R Kleiser, P Reubens; 140tr: The Picture Desk/The Art Archive/Gianni Dagli Orti, Musée Carnavalet, Paris; 140b: The Picture Desk/The Art Archive/Gianni Dagli Orti, Bibliothèque des Arts décoratifs, Paris; 141t: Leemage/Photo Josse, L-L Boilly, Musée Cognac-Jay, Paris; 141b: Leemage/AISA, Musée Carnavalet, Paris; 142t: Leemage/Photo Josse/G E Opitz, Musée Carnavalet, Paris; 142b: RMN/J-G Berizzi, Musée Adrien-Dubouché, Limoges; 143t: J-P Delagarde; 143b: Eyedea/Hoa-Qui/P Escudero; 144/145: AKG-Images/Palais du Belvédère, the Portrait Room, 1888, C Goebel, Kunshistorisches Museum, Vienna; 146bl, 146br: Leemage/Selva Collection; 146tr: Leemage/Photo Josse, private collection; 147l: The Picture Desk/Collection Daglie Orti/ Gianni Dagli Orti, Capitoline Museum, Rome; 147r: The Bridgeman Art Library/Barbara Singer private collection; 147b: Rue des Archives, The Granger Collection, New York, J Highmore; 148l: © Musée des Arts et Métiers-Cnam, Paris/Photo C Le Toquin; 148r: AKG-Images/ E Lessing, J-L David, Metropolitan Museum of Art, New York; 148b:The Picture Desk/The Art Archive/E Tweedy; 149tl: © Montres Breguet SA; 149bl: Corbis/L de Selva; 149br: Leemage/Costa; 150tl, 150b: Cosmos/SSPL/Science Museum; 150r: Corbis/Bettmann; 151bl: The Bridgeman Art Library/The Stapleton Collection; 151c: Leemage/ Costa; 151br: © Musée des Arts et Métiers-Cnam, Paris/Photo Studio Cnam; 152l, 152c, 154d: Cosmos/SSPL/Science Museum; 152/153b: National Army Museum, London; 153l: Leemage/ AISA British Museum, London; 1534: Leemage/ Photo Josse, G E Opitz, Musée Carnavalet, Paris.

Illustration of the Appert autoclave on page 88 by Jean-Benoît Héron.

THE ADVENTURE OF DISCOVERIES AND INVENTIONS
The Dawn of Industry – 1750 to 1810
is published by The Reader's Digest Association Limited,
11 Westferry Circus, Canary Wharf, London E14 4HE

The book was translated and adapted from *À l'Aube de la Révolution Industrielle*, part of a series entitled L'ÉPOPÉE DES DÉCOUVERTES ET DES INVENTIONS, created in France for Reader's Digest by BOOKMAKER and first published by Sélection du Reader's Digest, Paris, in 2010.

Translated from French by Peter Lewis

Series editor Christine Noble
Art editor Julie Bennett
Designer Martin Bennett
Consultant Ruth Binney
Proofreader Ron Pankhurst
Indexer Marie Lorimer

Colour origination Colour Systems Ltd, London
Printed and bound in China

READER'S DIGEST GENERAL BOOKS
Editorial director Julian Browne
Art director Anne-Marie Bulat
Managing editor Nina Hathway
Head of book development Sarah Bloxham
Picture resource manager Sarah Stewart-Richardson
Pre-press account manager Dean Russell
Product production manager Claudette Bramble
Production controller Sandra Fuller

We are committed to both the quality of our products and the service we provide to our customers. We value your comments, so please feel free to contact us on 08705 113366 or via our website at **www.readersdigest.co.uk**

If you have any comments or suggestions about the content of our books, you can email us at **gbeditorial@readersdigest.co.uk**

CONCEPT CODE: FR0104/IC/S
BOOK CODE: 642-005 UP0000-1
ISBN: 978-0-276-44517-0
ORACLE CODE: 356400005H.00.24